THE LANDSCAPING IDEAS OF JAYS

Judith Larner Lowry

UNIVERSITY OF CALIFORNIA PRESS BERKELEY LOS ANGELES LONDON

The Landscaping Ideas of Jays

A NATURAL HISTORY OF THE BACKYARD RESTORATION GARDEN

University of California Press, one of the most distinguished university presses in the United States, enriches lives around the world by advancing scholarship in the humanities, social sciences, and natural sciences. Its activities are supported by the UC Press Foundation and by philanthropic contributions from individuals and institutions. For more information, visit www.ucpress.edu.

University of California Press
Berkeley and Los Angeles, California

University of California Press, Ltd.
London, England

Line illustrations by Kathleen O'Neill

I wish to thank the editors of various publications for granting permission to reprint revised versions of my earlier essays. The following pieces have been incorporated into this book: "Birdsong Ripens Berries, Wind Brings the Seeds," *Orion* (May/June 2005), reprinted with the permission of Orion, 187 Main Street, Great Barrington, MA 01230, www.oriononline.org; "Gerda Isenberg's Contributions to Native Plant Horticulture," Rancho Santa Ana Botanic Garden, Occasional Publications 3 (1999), Claremont, CA; "The Gardener and the Quail: A Bolinas Love Story," *BayNature* 2, no. 2 (2002): 26–31; and "Caching In: The Landscaping Ideas of Jays," *BayNature* 5, no. 4 (2005): 20–24, 30. The book epigraph is from *Selected Poems: Revised,* by Czeslaw Milosz. Copyright © 1973 by Czeslaw Milosz. Reprinted by permission of HarperCollins Publishers.

Library of Congress Cataloging-in-Publication Data

Lowry, Judith Larner, 1945–
The landscaping ideas of jays : a natural history of the backyard restoration garden / Judith Larner Lowry.
p. cm.
Includes bibliographical references and index.
ISBN 978-0-520-24164-0 (cloth : alk. paper) — ISBN 978-0-520-24956-1 (pbk. : alk. paper)
1. Native plant gardening—California. 2. Backyard gardens—California. 3. Restoration ecology—California. I. Title. II. Title: Natural history of the backyard restoration garden.
SB439.24.C2L684 2007
635.9'5109794—dc22 2006036219

Manufactured in the United States of America

15 14 13 12 11 10 09 08 07
10 9 8 7 6 5 4 3 2 1

This book is printed on New Leaf EcoBook 50, a 100% recycled fiber of which 50% is de-inked post-consumer waste, processed chlorine-free. EcoBook 50 is acid-free and meets the minimum requirements of ANSI/ASTM D5634–01 (*Permanence of Paper*).

To my family

The publisher gratefully acknowledges the generous contribution to this book provided by the General Endowment Fund of the University of California Press Foundation

Do not die out, fire. Enter my dreams, love.
Be young forever, seasons of the earth.
Czeslaw Milosz, “Winter,” 1982

CONTENTS

SPRING

SUMMER

THE FIFTH SEASON

PREFACE

The seasons wake us up, turning a key in the heart that renews our view, sparks engagement in seasonally appropriate activities, and reinvigorates our lives. Where I live and work, on the north central coast of California, the activity of seed gathering most alerts me to seasonal change, with almost every month bringing many species to my attention for harvesting. Until we began holding workshops at Larner Seeds, this sequential ripening was my major seasonal marker. Once we opened our demonstration garden as an outdoor classroom, featuring guest lecturers, native plant edibles, common-naming contests, and celebratory songs, I learned that specific activities accrued to workshops held at certain times of the year, with certain questions and themes arising around these activities. Reveling in the seasonal manifestations of California's natural phenomena with the participation and liveliness of other similarly inclined people became a kind of play and celebration. With a common focus and shared concerns, we formed for those short times a temporary culture of the garden.

From these workshops I have taken the organization of this book, which hinges lightly upon the seasons of California. I add to the usual four a "fifth season," which is that time between summer and fall when no rain has fallen for three to five months and none is expected. In our semiarid climate, with wet winters and dry summers, the fifth season, when plants are in a holding pattern, produces a feeling unique to this region. Also perhaps unique to our state is the way that gardening in California can be forgiving, seasons sliding one into the other. Suggested activities appropriate to each season are just that—suggestions.

Some California Indian creation stories describe how Jay was given his work to do, planting acorns to grow into oak trees to feed the people, whose work was to hunt, harvest, and prepare acorn. At our first fall workshop, a western scrub-jay, in full view of thirty people, stole a wildflower-seed cookie from a plate, flew off with it, then returned for more. When I covered the plate, he grabbed an acorn from a bowl and made off with that too; I believe the tree that grew from it now shades the nursery. With this book, I honor the ideas and presence of that nervy jay, and the presence and participation of all the wild inhabitants of California, still here, still with us. May our gardens further their prospects.

My Dream Neighborhoods

California is the land of unfailing contrasts. Hot or cold, wet or dry, green or brown, low or high, you may order what you will, . . . and lo, it is yours within the hour.

William Dawson, *Birds of California*, 1923

No state possesses a more distinctive or multitudinous native flora than California. An unrecognized treasure, it should be ranked in equal value with our climate and our scenic beauties.

Lester Rowntree, "Wildflower Sanctuaries," 1934

Somebody shoulda gone to Sutro Heights and looked at Sutro Heights before Sutro got ahold of it. It was probably one of the great endemic areas in the Franciscan zone. . . . To me, it's extremely unappealing because I can't look at it without thinking: "What was there? What did grow there?" We'll never know.

James Roof, "The Franciscan Region," 1975

As I walk and drive through California, I mentally relandscape the world and hardly know that I'm doing it. Work on dozens of design projects has programmed my brain so that I almost unconsciously project images onto the current reality. In my mind, a landscape emerges where each hill, each flat, each slope regains some approximation of the unique vegetation that it once possessed, in a new arrangement that encompasses but is not overwhelmed by modern realities of house, car, and human population. I drop an oak grove here and a stand of willows there, reforest that patch with ghost pine, blue oak, and madrone, another with tanoak and redwood. I weave basketgrass meadows and hazelnut copses through the median strips, choose low-growing manzanitas to cling to steep banks, and fling dwarf coyote bush onto sea bluffs that might otherwise erode into the sea.

In my imagination, I weed the roadsides as I go, removing the invasive species that follow our cars and preventing further spread into the land. I employ a panoply of strategies, from mowing to plowing to mulching, and also a favorite fantasy: my backseat full of weed specialists, my Dream Team. Tools in hand, they leap out of the car at a moment's notice to remove the harbingers of ill tidings: the first Italian thistle, the lone clump of pampas grass, the ominous stand of yellow star thistle, and that ubiquitous quartet that follows the freeways from Monterey County to the Oregon border: poison hemlock, Harding grass, teasel, and fennel.

Driving and looking, I restore drainages, create native prairies, and people these imaginary spaces with children—swinging from tree branches, peeling bark, smelling wildflowers, and admiring vistas—who absorb through their pores the qualities of California's plants, seasons, and physical features.

Turning off the freeway into the neighborhoods of a city or town, I superimpose wildflower fields where now there are lawns, so that front yards become part of a coordinated effort to bring back something of the miles-long sweeps of annual spring wildflower bloom so remarked upon by early Californian observers. My eye does not stop at each lot line; nor do the pollinators, lizards, butterflies, or desert tortoises. Native bunchgrasses, whether in full flush of growth or semidormant, here set the pattern upon which the eye rests. A wide variety of annual and perennial wildflowers like Ithuriel's spear, elegant madia, or Indian paintbrush make their presence known through the year. Side yards are hedges of native shrubs for privacy and the fostering of bird and insect life. Native fruits, like huckleberry and woodland strawberry, coexist with their domesticated counterparts in vegetable gardens and orchards.

Backyards are painted with local native plant palettes, from coastal scrub to Joshua trees to redwood forest, which create livable, unique gardens that nurture and protect native fauna as well as shade and surround humans. These yards become laboratories that are beautiful, ever changing, and always interesting. I place benches made of local wood under oaks, under hollyleaf cherry, under cacti, under native fan palms. In this dream, native greens and fruits, including huckleberries, Indian potatoes, and camas, are featured at neighborhood potlucks, where the courage to try new ingredients, a notable feature of our modern cuisines, encompasses the culinary bounty of California as well. Camas pudding, brodiaea bulb slices in salads, hors d'oeuvres of peeled cattail stalks lightly steamed with wild onions are seasonal fare.

Native plant explorers in the wildlands are drawn ever onward, eager for the next plant event. What if a walk around the block provided some of that same sense of surprise? Species that once grew locally are the exotics now, novelties in their old territory. I imagine people walking through these neighborhoods with an air of astonished discovery, exclaiming "I didn't know that grew here!" or

"What is that? I've never seen that before!" In these neighborhoods, the health of the land is an accepted consideration, part of all decisions. Residents feel some degree of contentment from the cooperative, positive role they and their neighbors are playing, and so they feel less of a need to consume, to build trophy homes, to vacation far away, to move to a "better" place. Moving can be hard on the land and hard on us, making local experts an increasingly rare commodity.

THE NEW CALIFORNIA GARDENER

In some neighborhoods, my dream is not so far off, as a unique kind of gardening culture evolves in the San Francisco Bay Area and throughout California. With roots in a number of different gardening approaches, from organic farming to permaculture to ecological restoration, it has begun to create a gestalt of its own, which looks something like this:

From organic gardening it takes the goal of no chemical use. From xeriscaping it incorporates minimal water use. From permaculture it brings techniques like sheet mulching, retaining water on-site in bioswales rather than draining it off into the storm sewers, and the goal of continuous, cleverly managed food production without soil, air, and water degradation. From restoration ecology comes the desire to foster local flora and fauna. From resource management comes the idea of adaptive management, which involves reevaluating and adjusting strategies through time. From the indigenous peoples, along with deep respect for the qualities of the plants, come specific management techniques for continued renewal of health and vigor.

Combined, it's a heady brew, taking its unique shape in garden after garden, neighborhood after neighborhood. One characteristic of these new California gardens is that they function in multiple ways: as pleasurable landscapes, as significant food sources, and as restored habitats. Some turn into living labs, where "citizen scientists" can work on endlessly engaging connections presented at the back door, studying the survival of Pacific tree frogs in backyard ponds or using pollinator viewing in a wildflower field as an on-site contemplative practice.

Native plants are an integral part of many of these garden schemes, whether in a mixed flora of exotics and natives, in an all-native garden using horticulturally available species from throughout California, or, from the place where I most comfortably sit, in a garden focusing on the return of local species to territory once held by them. Here I have found what is for me the richest, most evocative story, the tale of the *backyard restoration gardener.*

This story has been a twenty-five-year-long venture in searching out what species might have grown here once and seeing what is possible in the light of their return, and also seeing what is not. The range of experiences includes surprises, miracles, intense frustrations and disappointments, calm gratification, ecstatic gratitude—all part of the daily bread for the backyard restoration gardener, whose prospects, motivations, and activities I present in the following pages.

Following this tale through the seasons, you will find sections on particular plants of interest at each time of year, from miner's lettuce in winter to clarkias in the summer, as well as activities that are seasonally based, like garden design in the fall, historical stories in winter, and renewal-pruning in the fifth season.

EVERY FIELD IS DIFFERENT

I have a friend who grew up on a dairy farm in central New York State. Driving around the fields and forests with him was always interesting. One early spring, noting no activity in the fields, I asked when the farmers would begin to plow. "Depends on the spring," he replied, and I remember the measured patience with which he added, "Depends on the field."

"All fields are not the same," he said. "Every field is different." He spoke as though conveying information that was so deeply known by him that it issued forth from his bones, as he listed the ways that fields could differ: by soils, exposure, slope, elevation, drainage, previous use, and more. Just out of high school, he had rejected a scholarship to a local agricultural college, where he feared they thought all fields were the same, or that the goal was to make them so—a more difficult prospect in the steep, rocky Taconic Mountains of New York State than in the Central Valley of California. He was what Wendell Berry calls "farm smart," an increasingly rare commodity now that the farming population is so small as to no longer be measured by the Census Bureau.

Somehow, my education lacked this great untaught lesson of site specificity, one that he had absorbed from childhood on—each field unique, every place different, every garden, every yard, one of a kind. The message for the gardener is the same as for the farmer; we too can note cues of ever-increasing subtlety as our interactions with a particular piece of land accumulate through the years. In California, with our great diversity of landforms and climate, and, consequently, vegetation, the motto "every field different" takes on even greater import than it does in other parts of the country. Our actions as gardeners and land managers can aid that which is so frequently praised—biodiversity.

The California Floristic Province, the territory where most Californians live, between the Pacific Ocean and the Sierra Nevada Mountain Range, is listed by Conservation International as one of the twenty-five most biodiverse, and most threatened, regions in the world. "In the last several decades, California has spent more money on conservation and set aside more habitat for protection than any other state in the United States. Nonetheless, the situation in California, the wealthiest of the United States, serves as an important reminder that biodiversity loss and the lack of complete and adequate protection for unique and threatened ecosystems is not just a problem in developing countries."

The heart of this book reflects my conviction that gardeners can be part of the reversal of California's status as one of the most degraded biodiversity hotspots in the world. Few states can be as well loved by their citizens as California, and few can be as replete with inhabitants who chose their home state at least partially because of its natural beauty. We have in some cases almost miraculously protected significant amounts of public land; we now begin to look at what the land managers called gardeners can do, from the tiny urban garden to larger acreages in less populous counties. We can begin to alleviate some of the threat to biodiversity through growing in our gardens local species of the many native genera that are found throughout California.

Examples of these uncommon but gardenworthy plants abound. Take a genus like *Phacelia,* for one, with eighty different species growing in different parts of California. Species from this genus are replete with virtues. Many are easy and beautiful garden subjects, make good cut flowers, are much frequented by native pollinators, reseed year after year, and add a delectable note of deep blue, blue purple, or lavender to the garden. Some will bloom in part and even full shade; others can take the glaring heat of all-day sun in the desert. Many have large, almost tropical-looking leaves, and a few are fragrant, a rather uncommon trait among California's annual wildflowers.

Similarly, there are sixty-one different species in the genus *Clarkia,* each with a different geographic range. Some botanists think there are as many as seventy different subspecies of our well-loved California poppy. Meadowfoam is another wildflower with localized species and subspecies, such as the Sebastopol meadowfoam *(Limnanthes vinculans).* As development proceeds apace and invasive species march across the land, habitat protection is the first order of the day. After that, gardeners can be key players in the protection of biodiversity, while enjoying floral pleasures seldom encountered in the gardening world, simply through inclusion of local native species.

Aldo Leopold, in articles written for farmers between 1938 and 1942, claims that conservation is not just a matter of preventing overuse but a matter of use that takes into account the ways that natural systems work. He calls our land-use activities awkward, a "disordering" just as much as a depletion, and in his piece "The Farmer as a Conservationist," it turns out that he too has a "dream neighborhood," a farmscape that weaves in the wild. He calls for farms with fencerows, with snags, with prairie remnants, with unchannelized creeks, marshes, and swimming holes. Perhaps Leopold could not then have imagined that our challenge would now be to hold on to these features in the face of development, that the job he claimed for the farmer would also become the job of the gardener.

I don't mean to imply that yards have no common aspects. The dream is not to make each yard different from every other yard—far from it. The patchwork effect of completely discontinuous landscaping visions is one with which we are all familiar. One of the best things that can happen, and I have seen this in my work with increasing frequency, occurs when neighbors get together to share intentions and, crossing borders, establish a bit of woodland, chaparral, or coastal scrub at their common boundary. Of course, since they live right next door to one another, the chances are good that their soils, aspect, exposure, and winds will be similar.

EVERYTHING'S HERE

I am sometimes asked how I can bear to limit myself to growing only what might have grown here before European land-use practices, the unintended incursions of invasive plants, and the ideas of gardeners dominated the terrain. I think of all the local species that are not yet part of my garden. In twenty-five years I haven't begun to exhaust the possibilities, and I am still surprised by what I find that is waning nearby that might make a welcome addition to the land around my house. The list of possibilities is long: from the harvest lily *(Brodiaea coronaria)* I glimpsed in the field across the street before the dirt bikers did away with it, to the native oatgrass *(Danthonia californica)* the gophers love too much (next time I'll protect it with gopher baskets). And the barberry *(Mahonia aquifolium).*

I recently saw it growing in coastal scrub up the trail, a few miles away but still in our watershed. Possibly its blue berries and scarlet leaves in fall may once have enlivened the plant palette hereabouts, yet there is no trace of it on the marine terrace where I live. I'm trying it out. On the same hill where I spotted the barberry, I found California aster *(Lessingia filaginifolia)* growing in bald, sparsely vegetated places sometimes called "barrens." The California aster has turned out to be a most satisfactory ground cover for the raised area around my pond. Its silvery leaves and attractive purple flowers have completely covered this infertile, potentially problematic mound composed of subsoil brought up from four feet below the surface during the construction of the pond. Happy to grow in such lean soil, it is appreciated by butterflies as a larval host plant and as a late-summer bloomer. Once there were salmonberries, fritillaries, lupines, lilies, clovers, clarkias, coast lotus here, things not now present, representing horticultural challenges enough for several lifetimes. It's not a matter of disliking non-native plants—of this I have been accused. Rather than a turning away from, it is a turning toward.

In some parts of my garden, I enjoy trying California native plants from other parts of the state, particularly the southern coastal areas, of which the Point Reyes Peninsula formerly was part. Usually they flourish, till we have one of our wetter or colder years. Then, they suffer, but survive. I am interested in the messages these plants send me about their limitations, and I enjoy their beauty, but for me the real juice comes from the return of what is local.

Now I have lived here twenty-five years, I still occasionally have to pinch myself to realize how surrounded I am by the plants that help me feel at home and that continually pique my interest and imagination. It seems too good to be true that this one acre—formerly covered with French broom, iceplant, Monterey pine, and cypress—now has, in addition to solid plantings of coastal scrub and coastal chaparral, coast live oaks that frame and shade my life. Willows and coyote bush hold sway, coast silk tassel and Pacific wax myrtle and coffeeberry give me bird and insect life to watch and sweet privacy to enjoy. Wildflower bouquets are mine for the gathering, to bring to weddings and funerals and birthdays; native berries and greens grow in the vegetable garden; and I have bay leaves enough for a thousand stews, hazelnuts to go with the morning's oatmeal, and an abundance of hazel wands for the fashioning of wreaths and trellises.

A friend of mine, Mary Bates Abbott, once brought her friend Milton "Bun" Lucas (Kashaya Pomo) to her forested land near Jenner. There he saw the oaks and huckleberries on the hillside, the willows by the creek, and the sedge used in basketry called "water gift" by the Pomo. She related, "As Bun looked at and walked over the land, which is a short distance from the Kashaya Reservation where he grew up, he kept repeating 'everything's here, everything's here,' as if he was seeing all his old friends again: plants, birds, the ancestral mushroom picking spot. My memory is of him looking around marvelling at how pristine and complete it all was for him. . . . It took him back to another time when life was more complete."

"Everything's here" became the name for the place, as in, "We're going up to 'Everything's Here.'" All that was necessary for survival, and for a full life, in the old way, was present to Bun's view. His reaction to that land reflects the broadest goals of the backyard restoration gardener, and ecological restoration in general, conveying the sense of peace and rightness that can come with this kind of gardening.

CROWNING EVERY HILLTOP

Sometimes, I spend time in my garden with no agenda or questions. I don't speculate on weeds or concoct strategies for their demise. I don't wonder why *Horkelia californica* is thriving here but not there; I just enjoy it where it is. Following birdsong mindlessly, I find that the peace of it is what matters. Other times, when I do go with questions, the answers may seep slowly into my bones, so that I end up with an idea of what to do without thinking about it much. I wanted a chance at something of the deep, half-conscious knowledge that my young farmer friend had been born to, and this chance my garden has offered. My place has schooled me.

Making our gardens the proving grounds for the restoration of public and private land can be a highly charged adventure or a relaxed exercise in patience. Research can be casual, focused, or open to the gift of grace. Such a gift was given to me once.

While getting ready to consult with a client in San Francisco, I looked through back issues of *The Four Seasons,* the occasional journal of the East Bay Regional Parks Botanic Garden in Berkeley. A 1966 interview with James Roof, first director of the garden, caught my eye. Roof spoke passionately about the beauties of the Franciscan flora, upon which we put a city—San Francisco. No quarter was given, almost nothing was spared. Roof in turn is unsparing in his critique of those who converted a previously treeless area to an unrecognizable mélange of species from all over the world. "They say, 'We made the bluff blossom.' I say, 'The bluff was blossoming before you got here, Jack!'"

In his interview, Roof gives us a compelling description of this unique flora before it largely disappeared. In those days, he tells us, the elegant shrub called blue blossom *(Ceanothus thyrsiflorus)* crowned every hill in the city. Though this coastal species usually prefers the lowlands, for some reason, in San Francisco, each hilltop wore the glossy leaves and the fragrant, sky blue blossoms of *Ceanothus thyrsiflorus.* A precious piece of the puzzle fits into place with this statement, site-specific raw material for my dream neighborhoods.

During my garden consultation in San Francisco the following week, I was asked by a very real client, garden spade in hand: "With what might we crown this hilltop?"

Fortuitously, I could answer, "Blue blossom."

The next day, he planted it. Dream come true.

Fall

1 *Esperando la Lluvia*

Waiting for Rain

Newcomers to California predictably experience difficulties grasping the unique realities of our climate. This is the land where moisture and heat are rarely partnered, and that, to paraphrase Robert Frost, makes all the difference. Garden books published in New York City or Boston or Chicago may not be particularly helpful here.

Nor does it help that the names of the seasons are the same here as elsewhere in the country, setting up expectations that our seasons will correspond to those in Minnesota, upstate New York, or Arizona, when perhaps only spring will be easily recognizable to the new transplant. We need new names!

Early explorers who arrived in California in the spring were generally delighted with the place, whereas explorers arriving in summer and early fall, before the rains came, were not impressed by the dry brown hills. For gardeners, fall is now well recognized as a favorable planting time, and so our seasonal round of the backyard restoration garden begins at this time, when the days shorten, the ground cools, and rains are possible. While the soil is still warm enough to promote root growth, rates of evapo-transpiration are slowed by cooling temperatures and ease the demands on the newly establishing root system; roots given moisture through irrigation or rain make slow and easy acquaintance with new territory.

We live, at this time of the year, with uncertainty. Fall is the time of waiting for rain, which translates into Spanish as *esperando la lluvia.* The verb *esperar,* with its double meaning of "to wait" and "to hope," is particularly appropriate for this renaming: *Esperamos la lluvia,* we are waiting, and hoping, for rain. Which Californians, especially those who have lived through drought, must always and forever do.

Smug is the gardener able to get plants in the ground or wildflower seeds sown just before a rain, knowing that everything is right for the easing of that plant into its new home, for the germination of seeds. The first rain requires acknowledgment and gratitude, as the great forces conspire to bring moisture back to the land, softening the soil and awakening dormant inhabitants of the rhizosphere.

And in the fall we pack seeds . . .

2 Birdsong Ripens Berries, Wind Brings the Seeds

> Plants reveal the hidden qualities of particular soils and the subtle properties of place and site. In airborne pollen, leaves, and winged seeds they give air visibility and demonstrate the mysteries of season and sequence.
>
> Paul Shepard, *The Others: How Animals Made Us Human,* 1996

In the fall, we pack seeds, working on a long, well-oiled table made from the rootstock of a native California walnut tree, onto which was grafted English walnut. Milled and planed, the figure in the wood reflects that history, being rich and dark at the rootstock end and lighter and creamier at the grafted end. On this shining surface, the three of us measure out, with spoons, cups, scales, and fingers, seeds of California native plants. We package them and make them ready for shipping to our customers—backyard restoration gardeners, volunteers working in public parks, botanic gardens around the world, academic researchers—to make more ceanothus, toyon, baby blue eyes, coast live oak. To make more purple needlegrass, goldfields, gumplant, and pink flowering currant. More wildflowers, more bunchgrasses, more native trees and shrubs. We join the ants, birds, wind, water, and other agents of dispersal to increase the presence of native plant communities where once they might have flourished, or where they grow now in a diminished state.

The table sits in front of a picture window, against which are propped lists of the birds and insects that we watch as we work. Our view is framed by California hazel, and in the distance we can see blue blossom ceanothus. To sit under blue blossom in the spring is to be surrounded by the buzzing, droning, whirring, and humming of insects of all kinds, especially native bees.

Now in the fall we observe the results of the work of those springtime pollinators—blue blossom's rich provisioning of a myriad of small land birds, many just returned from exhausting migratory journeys and in need of optimal nourishment. For weeks, the dull black, papery, three-chambered seed capsules split open to release gray black seeds to the chair cushions, sidewalk, and leaf litter below. From the cushions I brush the seeds into stainless steel bowls, then clean, dry, and package them. Retrieving the seeds from the litter is a task left to the rufous-sided towhee, the California thrasher, and the gold-crowned sparrow. I sweep the seeds and leaves off the sidewalk and by the next day, it is covered again with the detritus from their seed-finding mission. Such gardening tasks lead me to muse on the plants and other creatures that have coevolved in this place.

When I moved here to the northcentral California coast twenty-seven years ago, I was disappointed by the lack of intact habitat in the immediate environs. In my semirural neighborhood of fishermen, ceramicists, and financial planners, there was little for a native-plant seed collector to gather. I roamed restlessly north, south, and east, where human impact was less apparent, collecting as I went, intent on re-smothering every inch of my one acre with my best guess of what was once here, floristically speaking. To some of my ideas, the land said yes, to others, no. The intensely showy dune and bluff plants I found at the Point Reyes Lighthouse to the north did well here for several years, but the soil proved too rich for most of them, the land too flat, the drainage not quick enough. They were overtaken by our plantings of coastal scrub and chaparral shrubs like coyote bush, toyon, coffeeberry, and the beauteous blue blossom, which have proved livable, friendly, and sheltering.

Today my one acre of restored and garden land is in constant play with the wildlands. We bring in seeds from nearby native populations, including threatened herbaceous species like the Point Reyes checkerbloom, a glossy-leaved ground cover with large pink flowers, or the native-bee-friendly meadowfoam, staunchest of reseeders. We plant them where a solid phalanx of coastal chaparral and scrub plants will shield them from incursions by neighboring weed fields. We then harvest the seeds from these grow-outs and offer them to our customers. As a glorious by-product of all this, I live surrounded by an intensified, continually reimagined version of what the land might have been like once, when managed by California's indigenous peoples.

SEEDS AS SEASONAL MARKERS

Next to the walnut table is a bulletin board to which are affixed lists of seeds that need to be collected each season. The lists are compiled from what we see ripening in the native plant garden, from memory, from data recorded in looseleaf notebooks that date back twenty-seven years, and from notes scribbled on my collecting envelopes. On a shelf against the opposite wall, rows of these large brown envelopes are arranged alphabetically by the botanical name of the seed they contain. Written on each well-worn envelope are collection dates, locations, and also the amount of seed to be put in each packet, the quantity based on germination rates and the time involved in collection.

For some species, we return to the same spot every year, for some, every other year or every five. We are careful not to be too efficient, taking in most cases only a tiny percentage of the crop, 10 percent or less. Some collections involve permit-required wildland hikes, some are on private land where owners have given their consent; others take place here in my own backyard, or in the yards of my employees, and in the vacant lots, weed patches, and roadsides between us.

Seed collection is woven into the fabric of all of our lives, so that walking the dog, visiting a neighbor, driving on errands, the annual trip to the dentist at the same time every year become occasions for small gatherings. Between a friend's house and mine, California figwort *(Scrophularia californica)* sends long, lanky seed stalks up above blackberry brambles. In August, a visit with her is an opportunity to gently bend these generous seed producers over, so that a fine rain of black seed pours into a container. We've all become devoted gatherers and growers, never going anywhere without our brown envelopes. I watched our office manager become addicted to the process; then our graphic designer succumbed. Seeds have a power; once you feel it, they will not let you go.

Many of these sturdy, tan envelopes have been reused eight or ten times. Withstanding many years' hard use, thick and enduring, they remind me of *parfleches,* the rawhide carrying cases of the Plains tribes. A variety of stains betray which ones were used for harvesting berries: elderberries, thimbleberries, hairy honeysuckle, creamy white snowberries, glistening black coffeeberries, rich red toyon berries, deep blue black salal berries, manzanita berries, huckleberries, and salmonberries. Though the records of nine or ten different collections can be confusing and my employees understandably deplore the practice, I plead for the envelopes' reuse. Messy scribbles record thoughts I had while out collecting, personal notations of both my state of mind and the state of the plants.

Some envelopes contain fevered lists of species growing nearby. These little impromptu floras reflect a need to memorialize an unexpected abundance, or a miraculously intact assemblage of plant species. The lists somehow don't capture it, though, the rapture and relief at the survivals, the sense of their fragility, particularly under the onslaught of invasive plants, and even more frequently, surprise at the discovery in unlikely places of species whose habits and preferences I had thought I understood. My brief notes are shorthand for the ever-resurfacing recognition that I will never run out of surprises from the plant world.

One battered envelope records a flora full of exclamation marks. A tip from a friend sent me to a certain rocky, thin-soiled, dry road cut, to collect seed of western columbine *(Aquilegia formosa),* which I am accustomed to see growing in relatively rich, mixed-evergreen woodlands. Here it grew in plenteous masses, more abundantly than I had ever seen in one place before. Next to it, according to my envelope, grew a succulent plant called sea bluff lettuce *(Dudleya farinosa),* whose gray green fleshy leaves and bright orange flower stalks I usually find near crumbling coastal cliffs. I note my speculation on these unlikely neighbors: "An unexpected vein of water coming through the rock for the columbine, sharp drainage for the succulent dudleya."

Another envelope tells a tale of collection anxiety, where first I believe I have missed the harvest and then find that I have not. I was looking for the seed of vanilla grass *(Hierochloe occidentale),* which grows ten miles north of here. At the top of the wooded hill, the seed had already dropped to the forest floor, but grasses growing in openings midway down the hill contained ample ripe seed. Although I know this pattern, every year I forget, and, until I consult the notes on the envelope, begin to imagine the disappointed customer who will not be able to grow this unassuming plant with the intoxicating fragrance in its leaves and flowers. The envelope also tells me that two collections were made one month apart and that I particularly enjoyed the second collection date, when not only were more ripe vanilla grass seeds available, but also—an added inducement—evergreen huckleberries *(Vaccinium ovatum)* were ripening in the same Bishop pine woodland.

In his last manuscript, *Wild Fruits,* Henry David Thoreau described the ripening and biology of all the fruits he encountered in his extensive rambles in New England, but to the huckleberry he gave his greatest endorsement, asking: "Are they not the principal wild fruit?" Both the West Coast and the East Coast are rich in members of the genus *Vaccinium,* so Californians, Oregonians, and Washingtonians can join in Thoreau's enthusiasm. Since I live on a sandy marine terrace where these huckleberries usually don't grow, or grow sparsely, we must travel a bit to harvest these tiny berries, at once sweet and tart, so flavorful that the idea of missing them produces an uncomfortable sense of loss, of missed opportunity, of being wasteful of our prerogatives and privileges.

This feeling arises from deep, half-conscious roots, different from though probably connected to the fear of starvation. There's a sense of the possibly negative consequences of being an ungrateful guest at the feast. Many native people advise that plants respond to respectful, skillful harvesting and disappear without it. Cache Creek Pomo elder Mabel McKay said, "When people don't use the plants they get scarce. You must use them so they will come up again. All plants are like that. If they're not gathered from, or talked to and cared about, they'll die." Ethnobiologist M. Kat Anderson adds, "Today California Indians often refer to these practices as 'caring about' the plant or animal . . . caring for plants and animals in the California Indian sense meant establishing a deeply experiential and reciprocal relationship with them."

Many thoughts are in my mind; the fog coming and going on the ridge where I gather, the welcome sight of laden bushes, the immediate savor of eating the berries, their weight in the pail, the tiny seeds when cleaned and dried for packaging, and huckleberry raisins in winter breakfasts of oatmeal. I plan a huckleberry-gathering day, aiming for midharvest. Too early in the season and we spend precious time removing the green berries, too late and pickings will be slim, or the berries will have vanished for another whole year. Examining the vanilla grass growing in my garden, I see that about half of the seeds have fallen, so up north, where seeds usually ripen later, the seed stalks will still be full. According to the notes on my seed envelope, the huckleberry bushes will probably still be loaded also. In this way my backyard restoration garden functions as a calendar, informing when it is time to go out to wild land, which seed envelopes to bring, and when to clean the berry buckets.

PHENOLOGY

Using one ripening to predict information about another ripening was a half-conscious experience till I read the work of ethnobotanists Trevor Lantz and Nancy Turner, writing on phenology, the timing of life-cycle events. They describe in eloquent detail a system of seasonal reminders used by some British Columbian tribes. Natural phenomena that are relatively easy to experience, like flowers, like ripe berries, like birdsong, reliably occur at the same time as other important phenomena more difficult to observe, like roots ripening underground, like salmon leaving the ocean and beginning to move up the rivers and creeks. Plants and animals nearby, the so-called indicator species, tell of others farther away. As Paul Shepard says, "Phenology—the seasonal timing of life processes: sprouting, leafing, blooming, fruiting, quiescence—must surely have been one of the first sciences."

In this carefully recorded lore, events of great subtlety may signal when it is time to harvest vital resources. When Douglas fir cones shed the golden dust of their pollen, the chartreuse cambium of ponderosa pine that lies hidden beneath the outer bark is ready to remove for drying and then pounding into flour. The birth of mule deer fawns signals the ripening underground of avalanche lily bulbs.

THE SALMONBERRY BIRD

Some northwestern coast tribes connected the silvery, flutelike call of Swainson's thrush, which also rises up through our moist coastal forests farther south, with the ripening of salmonberries. The breeding call of *Hylocichla ustulata* was thought to awaken the berries and cause them to ripen. Mature salmonberries in turn indicated that the salmon were leaving the ocean to begin their freshwater journey from the mouth of the rivers and creeks inland; salmon harvest could begin. The bird we have named after English naturalist William Swainson was known to these tribes as "the salmonberry bird." This information inspires me to find a sunny spot in the garden in which to plant salmonberry, a species with which I am not very familiar, but which also grows in this region. *Rubus spectabilis* forms a brambly thicket and in the spring bears large showy red flowers on bare stems. I want to taste its raspberry-like fruits and see what birdsong sings them into ripeness here, further south.

Contemplating this profoundly different way of marking the seasons lets my mind float free to imagine, if only for one recharging moment, the freedom and the necessity to take in natural phenomena as life's main event. I marvel at the focus required to observe and connect synchronous happenings to each other, each pair signaling a time of efficient and concentrated harvest. "You young people," a Native American lady once observed to me some twenty-five years ago, "think we were able to do what we wanted to do whenever we felt like it. It wasn't like that." Indigenous peoples paid strict attention to certain life-cycle events, and the timing of their response was often critical to their well-being and survival. Though our lives are not so directly dependent on such timing, this work I am doing still requires that I get it right.

As layer upon layer of associated events accrue, we who work here build our own library of seasonal markers, which mix the botanical, the ornithological, the piscatorial, and the personal. I realize that my garden and seed room are replete with indicator species and information that can help me figure out some of what I need to know about my home. One seed envelope records a memorable collection day. Large-flowered fairy bell *(Disporum smithii),* low-growing lover of shady banks, grows in nearby coastal canyons. It was my birthday, and my daughter celebrated the day by collecting these large orange berries with me. She is a long-distance runner, and slowing down enough for this measured occupation was part of her birthday gift to me. Back home, she cleaned the berries as well, running them through the macerator with water, then pouring off the debris that rose to the top, while the clean triangular seeds, smooth and creamy as ivory, lay heavy in the bottom of the bowl.

Those seeds have long since been measured out, slipped into packets, labeled, and sold, but this envelope, still in use, brings back their cool feel and the pleasure of that day. This memory, replete with visceral experience and filial kindness (she was in high school then, finding her own path, and not much interested in mine) is an instance of my own version of phenological correspondences. The date of my birthday I now connect with the pumpkin orange berries of fairy bells, ripe on the one day of the year that my daughter does whatever I want to do, no matter how far from her own inclination. Maybe that leap out of adolescent egocentricity helped ripen the seeds. The beauty of these two life-cycle events, in any case, is now indelibly connected in my mind.

We are not the first mother and daughter to gather seeds together and to sense correspondences between events. The people who lived here before us, the Coast Miwok about whom we know way too little, and the Pomo, made their own phenological observations. In the early part of the twentieth century, a young anthropologist named Isabel Kelly spoke with two of them, Tom Smith and Maria Copa Frias. Her notes of those interviews are like sound bites, isolated fragments without context. We know more about the ethnobotany of the Pomo, in part thanks to a modern tome written by members of the Kashaya Pomo tribe, Jennie Goodrich, Claudia Lawson, and Vana Parrish Lawson, who explain, "The seeds [of grasses] were gathered in June or July when the first warm inland winds come to dry the grasses, causing them to throw their seeds. The women used to watch for these winds, knowing when they come there will only be a few days to gather the seeds before they fall to the ground."

Though spring is by and large an ecstatic experience here, there are many days when the wind rises like a monster in the late afternoon, buffeting, roaring, ripping tarps off woodpiles, plowing down fences that show any signs of weakness. Where trees are concerned, the wind culls the herd, bringing down dead limbs, uprooting those whose roots became too waterlogged during the winter rains to withstand its powerful ministrations. After big windstorms, I make hopeful drives through the town, looking for serious damage to the misplanted giants, the blue gum eucalyptus, the pines, and cypress. The low-growing, rounded trees and shrubs of coastal scrub, chaparral, and coastal forest absorb and are strengthened by the wind. For these natives the wind is an ally.

The wind is our ally too. It sings in our ear, "Go out, go out and collect the shiny, long-awned seeds of purple needlegrass." The wind knocks us on the head, saying, "Remember. Don't forget. Go soon or the seed will be all gone, ripped off the stalk by the blustering gusts."

So the year turns, and we gather, measure out, and package a small sample of its bounty and complexity to send to others who wish to live as much as possible with the plants, and their associated insects and animals, which belong to this place. Like the walnut of our work table, we have grafted ourselves onto the native, in the hope of creating a new kind of figure.

Sometimes I put everything else aside and try to empty my mind so that it can be filled by the great circling gyre; so that notice of time passing is given solely by the ripening of seeds, the emerging of insects, the movement of fish in the water, the maturing of roots underground, the appearance and disappearance of birds, the rising up and quieting down of their songs; so that the sweep of natural phenomena holds my attention long enough to connect events of importance. Like wind bringing down the seeds of native bunchgrasses whose roots go down fifteen feet, relatives of those same grasses whose seed was gathered one hundred fifty years ago by other women and their daughters. Like the bird whose song is so necessary and so powerful that it takes its name from the rich red fruit that ripens through the agency of its sound.

3 The Keynote Bird

Creating Habitat through Focus on a Single Species

On a pleasant fall morning in a neighborhood south of San Francisco, I pause while investigating a client's yard. Urban noises are noticeable but not overwhelming. Next door a baby cries, down the street a neighbor starts his car. In the yard western scrub-jays call, then a car alarm goes off and is quickly stifled. Mourning doves scratch in the litter under a hollyleaf cherry and a Meyer lemon tree, while fox sparrows cluster at the bird feeder, and a pale swallowtail careens through the yard, stopping to nectar at California asters in full bloom.

Sitting in this urban garden, I can see and understand some layers of the life here, while others remain elusive to me. The neighbor is going to work, and the baby is asking for food or attention, but what are the other creatures up to? Evaluating nectar resources? Looking for a mate? Breeding, nesting, and feeding babies? Some days it is enough to enjoy the beautiful flutterings and flittings, the sweet calls. At other times I hope for a more complete picture.

I have learned that a good way to deepen understanding is to choose one bird I see in my garden and use it as a way to enlarge my backyard musings. Making that bird a focus of my research, as I did with the western scrub-jay, and a theme for garden decisions, as I did with quail, increased my ecological literacy by leaps and bounds. The Point Reyes Bird Observatory (PRBO) Conservation Science also recommends using birds to guide us into habitat enhancement, noting that they are easily observed and respond quickly to changes in the environment, and "the important functions of birds in our environment cannot be overstated."

CHOOSING A KEYNOTE BIRD

At the ophthalmologist's office, I am given a form to fill out with a line asking in what way I would like my vision to be improved. How, it inquires, could my "lifestyle" be made better through their efforts? Optimistically, I request better long-distance vision for seeing birds far away. Like Keith Hansen—naturalist, guide, and bird artist—I would like to look up in the sky and identify without field glasses what is now merely a speck to me. Sadly, they cannot do this, so at least for now, I tend to focus on the larger birds such as the western scrub-jay and California quail, species that are easy for me to observe.

You might want to pick a bird that stays with you, a resident bird, one you can know at all stages of its life. Or perhaps you may choose a migrant that reliably appears in your garden, stimulating your imagination with its journey. It doesn't have to be a rare bird or one thrilling for its elusiveness. There is much to learn about the easily seen and easily recognized. Either way, by the end of a year with your keynote bird, new layers of meaning will accrue to your gardening activities and decisions.

Where to begin? To find out more about the western scrub-jay, I turned to the monograph on *Aphelocoma californica* in *The Birds of North America,* a valuable resource published by the Cornell Lab of Ornithology. Available at science or university libraries, these monographs summarize all the research that has been done on 716 bird species nesting in North America, a project completed in 2002 that is not only useful for ornithologists but also for anybody who likes to relax with birds at the end of the day or on weekends in their small or large, urban, suburban, semirural, or rural garden. The printed version and its electronic counterpart help us understand the details of interactions between birds and plants, many of which can be observed in the garden. There are worlds to be discovered here, presented accessibly. If you want to explore further, the bibliographies at the end of each monograph list citations going back as far as one hundred years of ornithological observation.

Curious about the availability of these useful publications, I sought them in several public institutions, without luck. I was unable to locate the monographs through the Mountain View Public Library, for example, but I did find an autographed copy of William Leon Dawson's four-volume classic, written in 1923, *The Birds of California.* Though in many ways dated, Dawson's book is an impressive endeavor and not to be missed. His purpose was not only to describe the avian species of California (to him "the most privileged task which might fall to the lot of an ornithologist") but to express "the very persons and lives of those hundreds of millions of our fellow travelers and sojourners. . . . To this end the birds have been viewed not alone through the rigid eye of science, but through the more roving, or tolerant, or even penetrating eye of the poet, the interpreter, the apologist—the mystic even."

This eighty-year-old classic contains descriptions of each bird's physical characteristics, habitat, and relationships, which preface lyrical and anecdotal passages so delightful, and frequently funny, that they led me to read about one species after the other. Dawson is a model of even-minded appreciation of all his bird characters, a characteristic frequently found in highly skilled ornithologists, who don't play favorites.

Modern field guides, with all their virtues, don't address the questions that Dawson asked in 1923, such as "Who owns the birds?" and "What are they good for, anyhow?" He came from a time when his "scientist, sportsman, poet" approach challenged him with its inherent contradictions, and when issues such as whether or not to kill and stuff specimens still troubled researchers and bird artists. These concerns contrast with the scrupulous care with which the PRBO Conservation Science's biologists presently seek to ensure that their observations are not harmful to their subjects.

Dawson tells us that, in the assemblage of his tour de force, in order to avoid preconceptions or stale data, he refrained from reading Audubon or Wilson. He has, however, helped himself liberally to information from the varied articles found in two ornithological journals that were published in the early twentieth century, *The Condor* and *The Auk.* The short articles therein range from casual observations made on an outing to more rigorous studies. Among these volumes, I found an unexpectedly lyrical and profound piece by Joseph Grinnell, "Uphill Planters," and a somewhat horrifying piece by Emmett Hooper on jay hunts. Scientific or university libraries are good places to find collections of these ornithological journals, which are frequently referenced in the bibliographies of the monographs contained within *The Birds of North America.*

ACORN DANCE

Drawn to the theme of birds that have an intricate relationship to acorns, I next investigated acorn woodpeckers. Though they are not present in my coastal garden, these beautiful jugglers may be seen one mile inland performing their carefully choreographed dance with the acorn. The bibliography from the monograph on the acorn woodpecker *(Melanerpes formicivorus)* sent me to a monograph written by Michael H. Macroberts and Barbara R. Macroberts in 1976. The Macrobertses carefully detailed every move in the complex ballet performed by the woodpecker, with the acorn as its passive partner. Marvels of avian choreography are recorded in their dedicated studies, revealing how thor-

oughly the nature of the acorn and of its storage unit, the tree trunk, are explored and exploited by the acorn woodpecker.

Unlike jays, who have been accused of purposefully leaving acorns on the ground to rot in order to harvest the worms that grow in them, acorn woodpeckers harvest directly from the tree. Opportunities for invasion—by weevils, fungus, and other entities—are greatly increased once acorns detach of their own accord and fall to the ground.

Harvesting directly from the tree, however, creates a dilemma for the acorn woodpecker, one which the Macrobertses describe with precision. The acorn forms the bird as the bird has formed the acorn. The shape of the acorn, as well as the way it hangs from the tree and its internal structure, are responsible for the details of this pas de deux. Acorns hang from oak trees with the blunt end attached to the acorn cap and the pointed end down. The acorn cap attaches to the twig. Of necessity, when plucking the acorn directly from the tree, the woodpecker grasps the pointed end in its beak, yet the storage holes it has carefully constructed in living or dead trees require that the pointed end be inserted first. Somehow the position of the nut needs to be reversed in the beak of the woodpecker, a challenging task for an animal without hands. This juggle is effected in several different ways, some of which were carefully photographed and documented by the Macrobertses. Sometimes the bird presses the blunt end of the acorn against a tree while the pointed end rests against its breast; the bird then slowly reverses the position of the nut until it finally holds the blunt end in its beak. This maneuver is reminiscent of the party game where a couple is required to pass an object held under one person's neck to the other person's neck without using hands or arms. Or the woodpecker brings the nut to an "anvil." This is a horizontal tree limb with cracks and crevices into which the acorn can be jammed so that it does not roll while the reversal is being made. The anvil can also be used as a place to crack open and eat some or all of the nut. Acorn caps are also removed here, perhaps because caps also present an opportunity for decay. Once the cap is removed, the other end, the blunt end, is grasped in the bird's beak. Only then can the acorn be inserted in the waiting storage hole, pointed end first.

The Macroberts focused their attention on the construction and maintenance of the storage holes as well. Matching the individual acorn to the specific storage hole is an art and a science. Holes are deepened or widened, debris is removed, acorns are removed and placed elsewhere if the fit is not tight enough, and the whole operation is given a care and attention that makes sense given the amount of nutrition available from a single acorn. I am reminded of the delicacy and precision with which acorn is prepared by the hands of native Californians.

Looking out my window, where a mixed flock of sparrows is enjoying a recent wildflower sowing, I consider my next investigation. Whether I pick up the monograph on wrentits, or on the spotted towhee, or on the gold-crowned sparrow, I know that interactions and relationships will be presented to me of which I now have no idea. The birds help me design my garden. Courtesy of my garden, I learn the birds.

4 I Live in a Quail Yard

A Garden Designed with Quail in Mind

Some of us living on the north central California coast can dimly remember metal signs affixed to barbed wire fences indicating areas that were once a quail refuge. The signs are gone now, but the California quail remain a feature in our lives. We don't know in what quantities our state bird formerly clucked and fussed, courted, nested, and raised chicks here when there were fewer of us, but on a foggy summer day, I can still raise winged thunder by walking out my back door. The male's assembly call still awakens us in fall and winter. And quail are still a familiar, beloved bird, seen rocketing at twenty-five miles per hour from garden to garden, rushing down the road on foot, or burbling half-hidden under coyote bush on a rainy day.

The desire to know them better and to cater to their requirements has become one of the design guides for my backyard gardening activities. When land management decisions need to be made, I frequently check with the quail. Their recent inclusion on the National Audubon Society's list of threatened bird species gives impetus to my efforts to provide for their needs.

California quail are among those creatures, like deer, bear, and coyote, whose wariness varies in accordance with a particular population's accustomed proximity to humans and status as prey. Though in the wild they are known and respected by hunters for their uncanny ability to disappear and their dislike of letting anyone within one hundred yards, in this semirural area, where they are no longer hunted, they are popular partly because of their visibility and the closeness they tolerate in garden situations. And their distinctive appearance, behavior, and vocalizations make them easy to identify. Binoculars aren't necessary to view their handsome chestnut, gray, black, and white markings and their chicken-like plump bodies and unique forward curling topknots, somewhat resembling a golf club.

In exchange for room in our gardens, they give the graceful gift of thriving among us. As they skim fences, ignoring property rights and heading for what they need regardless of who owns it, they stitch the neighborhoods together, providing a local totem and a topic of much conversation. Backyard coveys and garden flocks of *Callipepla californica* are a well-known phenomenon throughout California. Quail, like chickens, are gallinaceous birds, which means that observing chickens in the chicken yard can teach you much about quail, just as watching herds of goats can teach about deer. When allowed to forage in our wildflower plantings, they perform "the quail dance," a well-choreographed scratch scratch, jump back, peck peck, scratch scratch, jump back routine. A dance troupe of gray, white, and brown butterballs in constant movement reminds me of nothing but what it is, a moving mass of quailness, quail in abundance. It is but a remnant, though, of what used to be.

Before the advent of Europeans, quail played a part in the education of young Coast Miwoks, who were taught to shoot quail "in the brush—when it was raining, and the birds were wet." Several trapping techniques, including mile-long running fences with openings, were also employed by native hunters throughout California. For the Cahuilla in southern California, "Their favorite and most easily acquired game birds were the mountain and valley quail. Their wide distribution and terrestrial habits made them easy to capture. In the fall, birds in large coveys were easily captured by means of nets, traps, snares, and throwing sticks."

Quail topknot design elements, ubiquitous in California Indian baskets, speak to the importance of quail in native Californian diets. Gift baskets, the most precious and elaborate basketry types, frequently incorporated the actual topknots of quail into basket rims as well.

Jean-François Galaup de la Pérouse, the first European to describe California quail, said in 1786, "The brush country and the plains are covered with little crested gray partridges which, like those of Europe, live gregariously, but in coveys of three or four hundred. They are fat and of good flavor."

Walter Welch describes the coast of San Mateo County in 1867: "There I saw quail by the thousands everywhere; every canyon, gulch and ravine contained quail . . . and the whole country seemed to be alive with them."

J. Smeaton Chase, during his rambles up the coast of California in 1911, "wandered for an afternoon in Golden Gate Park, where I saw bands of quail running fearlessly among the shrubbery—a charming sight from which I argue that, notwithstanding the general belief, San Franciscans cannot after all be wholly bad." Yet in the year 2000 Golden Gate Park, which used to boast flocks of 8,000, counted only eight to twelve quail. The City of San Francisco, in recognition of the quail's appeal and drastically dwindling numbers, declared it the official city bird. Habitat restoration and feral cat control efforts are now under way in several parts of the city to bring local populations back from the brink of extinction.

With the arrival of Europeans and firearms, quail hunting became a major form of commerce and recreation. As favored game, quail were given preference in management practices. Here on the Point Reyes peninsula, before it was designated a national seashore, hawks were routinely destroyed to leave large numbers of quail and brush rabbits for human hunters. Jay shoots were initiated elsewhere for the same reason. Today, with quail still a popular target sometimes raised on game farms, hunters take close to half a million birds a year in California alone. With the spread of industrial agriculture and the advent of suburbia, quail encounter still more ways to die: hit by cars, crushed by plows, killed by dogs and cats.

Quail populations all over the state plummeted with the advent of the gold rush and its associated market hunting. New land use practices, such as plowing and overgrazing, along with the invasion of numerous less nutritious non-native plants, prominent among them the brome grasses, took a toll. Feral cats, called "tramp cats" by wildlife biologist and quail expert A. Starker Leopold, were implicated from the 1940s on in the demise of quail.

When commercial hunting and habitat destruction brought quail numbers dangerously low around the turn of the nineteenth century, measures to foster quail were initiated by local and state agencies. It is likely that the signs I saw long ago were a relic of a short-lived program of quail protection established in the 1920s between landowners and the Fish and Game Commission. Other steps included artificial restocking, strict regulation of sports hunting, the elimination of vast (in fact, horrendous) numbers of quail predators, and the banning of market hunting in 1880. When I witness the quail's impulsive rushes back and forth, their exuberant nervousness, and their consistent presence here, the word that comes to mind is "resilience."

Quail hunting, though anathema to some, is not, according to Leopold, nearly as destructive as the loss of the proper combination of dense cover for shelter, open ground for feeding, and low trees and tall shrubs for roosting. Some of the most eloquent writing about quail comes from hunters, such as Edward S. Spaulding in his book *The Quails,* written in 1949.

QUAIL AS PREY

Quail habitat should include a place to hide. Quail are everybody's dinner, and they lead a harrowing existence even here in this erstwhile refuge. They are eaten by hawks, by owls, by scrub-jays, by ravens, skunks, and bobcats, by foxes, by dogs, and of course, by cats. The depredations of feral cats on quail populations is a recurring theme in quail literature, and has been since the 1940s, when Spaulding wrote about the need to control cat predation. Cats can find quail in the densest stand of thorny brambles, pouncing on them before they can rise up through the network of tangled branches.

Quail may be likened to the California meadow vole, in that many species enjoy their succulent flesh, and the rise and fall of their populations can be of importance to a number of predators. In spite of the ominous increase in local extinctions, quail are still among the things of which there are many. Yet, as with the prairie dog, the passenger pigeon, and the buffalo, we cannot count on their populations remaining stable without our attention.

Thinking about quail habitat, I imagine this land before the Spanish brought thousands of cattle to graze year-round on the coastal prairies, before land was plowed up to grow artichokes and lettuce, and before gardeners appeared with their ideas of what the land should grow. One of the first acts by new residents when ranch lands were divided into small rectangular lots in the 1920s involved the planting of "windbreaks," using tree species from as far away as Australia (blue gum eucalyptus) and as near as Monterey (Monterey cypress and Monterey pine), trees that changed a windswept marine terrace from a complex mosaic of prairie, scrub, wetland, and evergreen woodland to an increasingly simplified pseudo-forest. Other invasive species, like French broom, pampas grass, capeweed, cape ivy, echium, and perhaps most important, non-native grasses, make the land increasingly difficult to recognize from the point of view of quail, wrentit, and saw-whet owls—or a gardener seeking to understand what makes her home unique.

WHAT DO QUAIL REALLY WANT?

The 1977 classic by A. Starker Leopold, *The California Quail,* is a guide to what quail require.

Food

Quail need open areas where the seeds and greens of native herbaceous species can be found. Leopold says, "The basic component of the quail diet is the assortment of seeds produced by various species of broad-leafed annual plants, botanically termed forbs" (51). In my garden, I grow annual, perennial, and shrub lupine species, including miniature lupine *(Lupinus bicolor),* sky lupine *(Lupinus nanus),* purple flowering shrub lupine *(Lupinus propinquus),* gully lupine *(Lupinus densiflorus),* arroyo lupine *(Lupinus succulentus),* and varicolored lupine *(Lupinus variicolor),* as well as red maids (*Calandrinia ciliata),* popcorn flower (*Plagiobothrys* spp.), and coast lotus *(Lotus formosissimus).* The quail's fondness for the seed and foliage of clover encourages me to focus on reintroducing this elusive component of the ecology of the coastal prairie. Leopold implicates the spread of weedy annual grasses in quail decline; we are down on hands and knees every spring keeping foxtail brome, velvet grass, African veldt-grass, and bluegrass out of the wildflower beds. We watch for subterranean clover, an introduced forage species that inhibits quail reproduction. Early spring greens, like miner's lettuce and native clovers, grow in beds for our consumption and to produce seeds, which the quail also enjoy.

Nesting

Though quail nests are notoriously difficult to find, quail are ground-nesters, availing themselves of a wide range of opportunities. After finding quail eggs in a clump of California fescue *(Festuca californica),* we increased our plantings of that handsome bunchgrass. Quail here also favor the base of one of the largest native bunchgrasses, Pacific reedgrass *(Calamagrostis nutkaensis).* Inland gardeners might grow deergrass (*Muhlenbergia rigens*) for the same purpose. Our prunings continually provide material for brush piles, which are left undisturbed during nesting season, April to July.

Cover

Quail require shrub islands for shade, shelter, and protection from their numerous predators. Our keynote plant, coyote bush *(Baccharis pilularis),* provides shade, peace of mind, and a good place to congregate on a rainy winter day. In one part of the garden, five or six coyote bushes, mixed with coffeeberry and blue blossom, grow thick and full, forming an island I call "Quail Village." Walking by this bosque, on a drippy day, I hear quiet burblings within its dense foliage and peer inside to see the quail strolling around underneath, safe and fairly dry. Here they also retire after the morning feeding to relax in what Leopold calls "daytime loafing cover," which he designates "the key building block of quail habitat" (Leopold, 49).

Large shrubs and small, dense trees, like blue blossom, red elderberry, bay, salt bush, and juniper, provide roosts for quail, who prefer to spend their nights off the ground. In my garden, a fifteen-year-old broadly rounded coast live oak next to the house is the favorite. The quail rocket through the twilight, filling all the spaces in the young tree, adjusting themselves for the evening. Open windows in the house allow the small sounds of this settling-in to penetrate the sleep of nearby humans, a rare and exquisite lullaby.

Dust Baths

The quail take note of and appreciate our gardening activities. In late summer, we use a weed whip to prune islands of native bunchgrasses and perennials, of Douglas iris, red fescue, dune tansy, and California figwort. The day after such an event we find the flock indulging in luxurious dust baths where we exposed the soil, sinking one at a time into a quail-shaped tub. Taking turns in a civilized manner, they exhibit an organized enjoyment of these bathing opportunities, reminiscent of portly bathers at a European spa.

Liquid Refreshment

A small pond or other permanent source of water is helpful for quail, though on the coast they can obtain a significant part of their liquid needs through fog, dew, and succulent greens. Inland quail need rivers, creeks, stock or sag ponds, or seeps, especially in the summer. We provide water consistently throughout the year in a small pond, planted so as to provide both open and sheltered drinking areas.

QUAIL BEHAVIOR

Once in midspring, quail dating season here, falling asleep by this little pond, I woke to find a quail's eye fixed intently upon me. He and his sweetheart were perched on a flat rock at the other end of the pond. He sipped from the pond, then watched his sweetheart while she drank, his murmurous, reassuring noises the background to which I fell back asleep.

I was charmed by this romantic scene. How they must love each other, the sociable quail. Speculating about quail behavior, I find it difficult not to anthropomorphize. Like us, at times they crave each other's company, then at other times, they don't. Leopold says, "When quail are flushed and scattered, their main preoccupation following the disturbance is to reassemble with their companions" (67). However, as soon as sex hormones are activated in spring by a combination of warmth and increasing day length, aggressiveness and irritability take over, pairs form, and the covey disintegrates, a dynamic not unknown to humans as well.

In their domestic lives, quail exhibit clear gender roles. Males take turns watching over feeding coveys from low trees and shrubs, roofs, and fence posts. By sharing sentry duty, they each have more opportunity to eat. When paired off in the spring, the males watch over their mates as they build nests, feed, and drink. Females lay clutches of ten to seventeen eggs, and if the first clutch of the season doesn't hatch, and sometimes even if it does, she is off to try for a second batch, not necessarily with the same guy. Socially inclined, as we are, they are also, like us, in a hurry, careening speedily from cover to open feeding areas and back again. The blur of their running legs is comedic, making it almost a disappointment when they finally take to the air, which they do with seeming reluctance and only when pressed.

These gamy birds, known in wild territory for their rocket-like speed and their uncanny ability to hide, adapt to a friendlier situation with alacrity, allowing humans to approach as close as ten feet before flushing. Some quail have even been raised as pets. One member of a related species, the northern quail, also called bobwhite *(Colinus virginianus),* inspired the well-loved book *That Quail, Robert,* written in 1966 by Margaret Stanger. Robert hatches from an egg found in an abandoned nest in a Cape Cod town and is lovingly reared by his human family.

The average age of a quail in the wild is one year and seven months, but Robert, who turned out to be a female, lived almost five years. When, after a long, remarkable life, she died, her family and friends were bereft. She was so much smarter than anyone would have thought, and so enjoyable to live with. There are also quite a few records of California quail raised as pets, though none, it seems, developed the rich personality of that quail, Robert.

The desire to get quail to "come close" is one experienced not only by hunters. One of my favorite quail stories is told by Carobeth Laird in her book *Mirror and Pattern,* written to honor and record the knowledge of her Chemehuevi husband, George Laird. George had a plan to entice the quail to come close; to that end he scattered seed all over his body when he knew they were near. Carobeth describes the scene.

> Soon the whole flock was climbing all over George, fluttering their minute wings to balance themselves on the uneven surface of his body, feeding eventually from his hands and face, pecking at his hair. All the while, George lay perfectly still, seeming not to breathe, until, in the gathering dusk, the mother led her drowsy brood back into the brush. . . . This was the type of experience that he would remember and cherish all his life.

Somewhere in the middle of the continuum from wild to domestic are the quail relations of my neighbors, who have maintained a feeding and watering station in their yard for twenty-five years. Their setup entails a wooden table under a large manzanita, where seed is regularly and artistically flung. At around 4:30 in the afternoon, a river of quail begins to flow from their loafing site to the feeding table. I am grateful for my neighbors' knowledge and their consistent observations of quail through the years. With their supplemental feedings, they may have saved our local populations by keeping them going through difficult times.

Resident here, the quail are born and die in our view. In the garden, they mold us back to an older contour, to the scheme of open prairie rich with herbs, to dense coastal scrub, and protective evergreen woodlands. Within these quail-appropriate spaces, many other species may flourish as well.

Some have it the quail assembly call says *cu-ca-cow* or *chi-ca-go,* but I prefer the early Spanish Californians' rendering, *cui-da-do.* Translated "take care," it is a more appropriate rendering for these apprehensive birds. The *Sibley Guide to Birds* syllabifies it as *put-way-doo,* and the Coast Miwok, who were here before us, said *tso-ko-tok* or *hek-hek-ki.* When the local preschool children come to visit the garden, they announce their arrival by imitating this call at the gate. Their voices don't sound all that different from the real thing.

At the end of a conversation with my quail-feeding neighbors, we toast the birds in the many syllables attributed to the California quail by California's native peoples:

To the quail. *Cu-ca-cow, qua-quer'go, chi-ca-go, ka-kai-ich, ka-ka-chu, tso-ko-tok, hek-hek-ki.*

May you flourish. May you grow and be abundant. *Cuidado!*

5 The Keynote Plant

Rules of Thumb for Garden Design

> A wild garden needs shrubs to show off the smaller plants, to give the garden a skeleton or a frame. It needs them for the settings they provide, the contrasts they make. They hold the garden down, and give it solidity and permanence.
>
> Lester Rowntree, *Flowering Shrubs of California,* 1939

As I move through my dream neighborhoods, I note that a visual calm emerges, an aesthetic harmony created by the repetition of one or several shrubs or trees. I think of these elements as *keynote plants,* chosen to make a background against which other elements play. The use of keynote plants, as well as making good design sense, furthers the theme of a deep knowing of plants and place. They may assume a totemic aspect, finding their way onto embroidered pillowcases, garden signs, and garden art. Just offhand, I know of an Oak Mesa, a Toyon Ranch, a Sagebrush Cabin, and our own Coyote Bush House.

Keynote plants differ from specimen plants, which are frequently chosen for their unusual characteristics, such as showy bloom or intense fall color, that are intended to catch the eye. Keynote plants usually are more unassuming, with perhaps a short season of flowering and a long season of solid foliage presentation. Outside of forested areas, they are usually shrubs. In an oak or madrone woodland or in a fir, redwood, or cedar forest, the trees are the keynote plants. They make a regular, tranquil statement above and offer much freedom below and between them for the planting of understory shrubs, grasses, annuals, and perennials. A continuous planting of a native grass, such as California fescue *(Festuca californica),* under black oaks or between Douglas firs can provide a harmonious background for various colorful herbaceous species, like Indian pink *(Silene californica),* woodland iris *(Iris macrosiphon),* or Ithuriel's spear *(Triteleia laxa).*

At Larner Seeds, as I've mentioned, coyote bush is our keynote plant. In the beginning, when I partially accepted the prevailing prejudice that coyote bush was an uninteresting junk plant that should be removed as quickly as possible,

I thought mine was a daring choice. Since then, more times than I can count, I have had the following conversation with local people.

"What's that full, shapely, useful, low-maintenance, drought-tolerant, evergreen shrub over there?"

"That? It's coyote bush!" A moment of surprised silence usually follows.

I understand their surprise. I admit that coyote bush looks significantly better in this garden than it does in some of the vacant lots hereabouts. The predilection to thrive here may reflect our horticultural practices, a particularly amenable location, or simply that it feels appreciated. Many native plant gardeners have learned the joy of taking a species that may look ragged and unprepossessing in the wild and giving it just the amount of attention needed to transform it into a tidy garden plant, one at which the most rigorous "white-glove gardener" will not scoff. A variation of this pleasure is to take a species that receives scant attention in the wild, one that never appears on the list of "California's ten choicest perennials or shrubs," and mass it or place it so that its attributes are set off—prompting visitors to ask, "What is that handsomely foliaged planting with the intriguing red flowers?" (It's California figwort, *Scrophularia californica,* seldom if ever mentioned in native plant garden books.)

Or, "What's that shapely, sturdy plant with the daisylike bright yellow flowers?" (This one is gumplant, *Grindelia stricta,* from a genus common in the wild, rare in the garden.)

Or, in the fall, "What is that very showy shrub completely covered with fluffy snowy white flowers that the insects love?" (Again, the answer is coyote bush.)

THREE ASPECTS OF GARDEN DESIGN

A useful way of thinking about garden design is to classify plants according to three variables: color, form, and texture. *Color* in this context refers to foliage rather than bloom and can range from deep olive gray through rich pure green to light chartreuse. *Form* can be rounded, mounding, erect, columnar, triangular, sprawling, or creeping. *Texture* is perhaps the hardest to define; it usually refers to leaf size, ranging from fine to medium to coarse. A plant with tiny leaves is generally considered fine-textured, but the relative rigidity or flexibility of the leaves is also taken into account. Foliage that flutters in the breeze, such as that of poplar, cottonwood, or aspen, has a finer texture than foliage so stiff as to scarcely acknowledge wind going through, such as that of manzanita or coast live oak. Plants with medium texture include coffeeberry, Pacific wax myrtle, and toyon. Coarse-textured plants range from those with huge, stiff, succulent leaves, like yucca, to those with flexible leaves at least half a foot long, such as California aralia, white sage *(Salvia apiana),* and western sword fern.

To create a satisfying balance between repetition and variation, I frequently apply the following rule of thumb: plants viewed together in a design should vary in no more than one or two of these three aspects. I usually aim for at least one aspect in which they are the same. With coyote bush as our keynote plant, I can add Pacific wax myrtle, which has leaves of a similar color but a coarser (read "bigger") texture, though their flexibility makes them closer to the less flexible but smaller foliage of coyote bush. If Pacific wax myrtle is given room to grow, it spreads out to become a large shrub or small tree that is almost as broad as it is tall, a form that coyote bush shares. When crowded, the wax myrtle becomes almost columnar.

If we follow this guideline, even loosely, we avoid adding to the sum of the visually painful experiences so prevalent in California gardens, where the climate doesn't constrain choice as much as it does in other parts of the country. In a not-atypical central California neighborhood, adjacent plants may vary in all three aspects. The coarse-textured, gray green, fountainlike yucca is found right next to fine-textured, deep green, mounding juniper next to a date palm next to a lemon tree next to a Colorado blue spruce. Each plant is essentially functioning as a specimen plant, set off by its neighbors and providing no restful, harmonious interludes for the eye.

EASY WAYS TO GET IT RIGHT

This lightly held rule of thumb should not create anxiety in the gardener. One easy and obvious way to get it right is to design with plants from the same plant communities. Automatic compliance results. Plants adapted to the same climactic and soil conditions are likely to have developed similar strategies for handling them; some manzanitas, for example, have tiny, waxy leaves to evade drought by reducing surface evaporation, while bigleaf maple, wild grape, and thimbleberry have large, flexible leaves to catch available light in wooded situations. Willows and alders, sometimes found growing together where the water table is high, are similar in form, texture, and color while varying in height.

Another neat trick to bring both variation and unity to a garden is to plant different subspecies or cultivars of the same species near each other. We grow several cultivars of coyote bush with prostrate habits, including *Baccharis pilularis* subsp. *pilularis* 'Twin Peaks' and *B. pilularis* subsp. *pilularis* 'Pigeon Point', as well as the tall shrub form, *B. pilularis* subsp. *consanguinea* (much less frequently grown in gardens), and a nice intermediate form with tiny leaves discovered on a nearby vacant lot that we call *B. pilularis* subsp. *pilularis* 'Bolinas'. Though they have the same form and foliage color, they vary by height and only somewhat by foliage size. On that same lot, we found three subspecies of coyote bush growing cheek by jowl, and we mimicked that combination in the garden. By combining similar though not identical foliage and form with sharply differing heights, a pleasing "estate-garden" feeling can be created.

If you want to use one of the prostrate forms of *Baccharis pilularis* as a ground cover, consider backing it up with the shrub form, *B. pilularis* subsp. *consanguinea,* as one of your background plants, perhaps combined with other coastal scrub associates. With that much coincidence of form, texture, and color, you could easily get away with introducing evergreen barberry *(Mahonia aquifolium),* with its contrasting color, texture, and form. Or consider interweaving different subspecies and cultivars of coyote bush with buckwheats (*Eriogonum* spp.), mugwort *(Artemisia douglasiana),* and California sagebrush *(Artemisia californica)* for a green and silver garden.

A similar scheme might be employed with the tall shrub coffeeberry *(Rhamnus californica)* interplanted with *R. californica* 'Eve Case' or *R. californica* 'Mound San Bruno'. The genus *Ceanothus,* with its many low-growing small shrubs and small tree species and cultivars, also offers this opportunity for creating combinations of plants with strong similarities and interesting differences.

CHOOSING THE KEYNOTE PLANT

For ideas, haunt local parks, refuges, and recreation areas, as well as botanic gardens and local native plant gardens. Also scrutinize the pages of the encyclopedic compendium by Carol Bornstein, David Fross, and Bart O'Brien, *California Native Plants for the Garden,* a supremely useful and carefully written tome that indicates that native plant gardening has finally come of age.

In your garden, your keynote species might be buckbrush *(Ceanothus integerrimus),* mountain mahogany *(Cercocarpus betuloides),* coast live oak *(Quercus agrifolia),* or black sage *(Salvia mellifera).* Each has its own mythology, its own aficionados—both human and "more than human"—and its own life history.

For your garden, it makes a place to begin. Using another helpful book, *A Manual of California Vegetation,* by John Sawyer and Todd Keeler-Wolf, you learn what other plants grow in the wild with your keynote plant. Black sage, for example, is part of twenty-three different plant "series," or associations, that include groups dominated by bigpod ceanothus *(Ceanothus megacarpus),* bigberry manzanita *(Arctostaphylos glauca),* and coast live oak. In these lists of associates, you will find possibilities to consider for inclusion in your garden.

If I lived on Catalina Island or the nearby coast, I might use Catalina currant *(Ribes viburnifolium),* which is evergreen and malleable, with a good dense habit and pleasant though subtle flowers that are appreciated by hummingbirds. The scale is right for the small garden. A Modesto gardener whose home is situated on what was once the floodplain of the Tuolumne River uses the yellow currant *(Ribes aureum).* Its fragrant yellow flowers light up the spring, looking particularly good against gray walls or weathered fences. Though it is deciduous, its leaves are gone for a very short time. In many parts of California, from the Santa Clara Valley to the valleys around San Diego, the lovely shrub called islay, or hollyleaf cherry *(Prunus ilicifolia),* meets the important keynote characteristics of being a good background plant, fast-growing, sturdy, and reliable, with beautiful fragrant white flowers and a strong indigenous lore based on its invaluable purple fruit.

In classic gardener mode, wanting what I can't have, I sometimes feel wistful about my inability to grow most manzanita species here in the fog. If I lived in one of the many places in California where manzanita species flourish, surely they would be one of my choices for a keynote plant. Handsome all year round, they belong to fairyland when their pink or white bell-like blossoms appear, and when their beautiful fruit ripens, they call for instant rounds of manzanita cider, at once sweet and tart. (I have included the recipe in the endnotes.)

In the hotter, dryer parts of California, I might choose palo verde *(Cercidium floridum),* in spite of its habit of losing its leaves and continuing without them through much of the year. Through repetition in the garden, dormant palo verdes make a good background for other plants, the light shade created by their rich yellow green branches making them a generous host for other desert species. For variation of form, some might be pruned into trees, others left to form a thicket. Your knowledge of this plant's place in the world will deepen in the spring, when its spectacular bloom of golden yellow flowers draws a host of insects. Desert willow *(Chilopsis linearis),* with its spring bloom of large, usually pink flowers, is another highly recommended shrub or small tree from the California desert.

TWENTY-FIVE YEARS WITH MY KEYNOTE PLANT

What seemed a daring choice for me twenty-five years ago now seems an obvious one, as the healthy green globes of coyote bush, symmetrical and neat through mulching and pruning, tie the garden together, make hospitable nooks for herbaceous species, and foster the insect and bird life that coevolved with it. We can prune out lower branches to reveal the surprisingly interesting sculptural quality of its alligator-hide branches or cut the whole shrub to the ground to begin anew, as fire would once have done. Through my years of coyote bush observation, I've learned that, like mesquite in the desert, coyote bush functions differently in different places. Here it might be part of the coastal scrub; elsewhere, it might be transitional to oak forest. One publication says that it is flammable and should be avoided; another says it is a good fire retardant when given some irrigation and kept pruned. Coyote bush is reliable, unassuming, and well known, while at the same time it is ambiguous, complex, and unpredictable—all good characteristics for a keynote plant.

When customers ask what would be the equivalent of coyote bush for their garden, I am unsure. Though I can make some good guesses, I can't say for certain what would be as easy and useful in other areas as coyote bush is here. However, since I didn't know twenty-five years ago just how easy and useful coyote bush could be, I can only assume that other equally surprising keynote plants await garden use in other parts of California. Your keynote plant may also be a familiar species, about which new worlds will be revealed through your use of it in the garden. Who knows what else I'll be learning about *Baccharis* in the next twenty-five years, as the common continues to prove uncommonly interesting.

6 Sing Willow

> The willow (genus *Salix*) has always been extremely popular for place naming because its presence usually denotes running water. In California, there are about twenty species, widely distributed in the state. The maps show more than two hundred Willow Creeks, and more than a hundred Willow Springs, Sloughs, Lakes, and Valleys.
>
> Erwin G. Gudde, *California Place Names,* 1998

Driving south between the towns of Tomales and Point Reyes Station in mid-November, I am watching willows. Each fold in the dry, dry hills carries seasonal creeks and seeps that form tiny wetlands before they make their way either directly to Tomales Bay or first to Walker Creek, which drains into Tomales Bay. Each crease in the land is held by willows.

My eyes find rest in the leafy thickets, which, even this late in the year, are the quintessence of "verdant." The species found along this road, arroyo willow *(Salix lasiolepis),* grows all over the West, reaching heights of from eight to twenty-five feet, with an equal width. It can be found as far north as Alaska, east through South Dakota, and south to New Mexico. Here in California it is almost Thanksgiving, yet the arroyo willow is still holding onto its long, narrow, olive green leaves. Most of the groves are still green, though some, in the drier drainages, are beginning to pale into gold.

IMBIBING WILLOW

At home, while thinking about the landscape uses of native willow, I boil up and drink a mild decoction of its bark and leaves. I toss a handful of willow material, including leaves and twigs, into a stainless steel pot filled with water, which I boil for a while and then strain. I imbibe it in the tradition of William Least Heat-Moon, who, as part of his research on prairie plants, performed a series of courageous botanical experiments.

> I've eaten parts of some and brewed several extracts according to old receipts: one of them I couldn't bring myself to swallow (a writer most fails when he loses nerve) but another I drank two cups of. Soon I lost my legs and had to lie down, and my mind seemed to tumble as if old tethers broke, and I dozed off into some grassland dream, and when I woke all I brought out of it was a recollection of an entanglement of long and numerous yet nearly invisible tendrils of a radiating vine; also in my head were the first six words of this chapter.

I expect no such dramatic consequences from drinking this mild willow tea, which has been used for so long (at least four thousand years), in so many parts of the world (including Asia, Europe, Africa, and North America), and by so many that its effects are relatively well known. Though contraindicated for some with allergies, asthma, blood conditions, and digestive problems, this pleasant brew unites me with a long, mostly positive history of plant-human interaction. Employed to quiet the agitation of arthritis, to lower fevers, control diarrhea and chills, heal laryngitis and lumbago, and soothe the pains of childbirth, willow has also been applied as a compress to insect stings and bites, poison oak rashes, and minor burns, and willow infusions have been used as an antiseptic wash for skin problems of all kinds. The leaves and bark are rich in vitamin C and salicin, an intermediate form of salicylic acid. Salicylates have been artificially manufactured for the last one hundred years to produce the most famous pharmaceutical compound in the world, aspirin. In the last century, over one trillion aspirin tablets were sold.

The most frequently touted herbal form of salicylic acid comes from the dried bark of white willow *(Salix alba).* Colonists arriving in North America with whips of *Salix alba* under their arms noted with surprise that the indigenous people were also using native willows for similar purposes. They did not, however, investigate local sources, preferring, in the initial stages of a long history of ignoring indigenous flora, to use and propagate the European species. Consequently, white willow has now naturalized throughout the eastern part of North America. Many of the "willowists" in this country, those who promote and study the use of willows for medicine and basketry, seem unaware that we have our own species in this country, though native willows can be found in every state in the union.

WILLOWS AND WILDLIFE

Present in varying degrees in different species, the salicylates in willows are secondary plant chemicals that protect members of the genus *Salix* (as well as other members of the family Salicaceae) from bacterial, fungal, and viral invasions. Frequently toxic or repellent, these chemicals also protect against mammalian herbivory.

At the same time, insect herbivores use them to repel their own predators. Leaf beetles munching on willows that are high in salicylates do better than larvae munching on willows that are low in salicylates. Many leaf beetle species in the larval stage secrete a repellent-smelling secretion when disturbed by enemies. If the willow of their choice is high in salicylates, they will secrete more of this bitter-smelling substance, increasing their chances of survival. In some cases, these salicin-rich clones showed more defoliation than the clones growing in a similar environment but with less salicin. The presence of salicin may have at one and the same time benefits and detriments in preventing herbivory for willows.

But willows can take a lot. Christopher Newsholme, the author of *The Genus Salix,* asserts that "it is very exceptional that any permanent damage is caused by the numerous and varied assortment of aphids, caterpillars, willow beetles, and weevils that all participate actively in the ecology of the willow."

During rutting season, late August, the local deer gouge long vertical wounds in my garden willows with their antlers. Bark hangs in tatters from these deep gashes, which occasionally show traces of deer blood as well. Willows are singled out for this treatment; no other species is touched. Since the shrubs are not girdled, and because willows have such exuberant regenerative powers, these actions are not fatal. Even willows growing in containers as nursery stock are used in this rough manner. Why this marked preference for willows? I wonder if some soothing ingredient is sought. Or maybe the texture is just right for making the loud noises required by male deer at mating season.

Birds also must rejoice in the sight of willows, knowing that they may find shelter and food in their secluded fastnesses. Should I venture into these tangled masses, I might startle an inordinate number of birds, both resident and migratory. They may be nesting, resting, or gleaning. Birders go to willows to find, among others, Wilson's warblers, chestnut-backed chickadees, willow flycatchers, California quail, willow warblers, and western scrub-jays, and once upon a time (1927) but no longer, quantities of the yellow-breasted chat. Willow buds are important winter food for birds, and the galls that form on willows supply larvae for protein.

Willows and alders, often found growing together, make a matched pair, willows providing dense nesting niches around three feet off the ground, while the taller alders provide song perches. In the winter, leafless, these two present a harmonious picture, similar enough to "go" together, different enough for interest. For the larger garden, planting alders (to sixty feet tall) in the background and willows (to thirty feet tall) in the foreground guarantees a quick-growing spectacular planting.

WILLOW AS PRIMAL MATERIAL

The growth habits of willow, its ability to resprout when cut to the ground, make it a symbol of death and regeneration to many peoples. That which seems to be dead, lives, that which seems to be gone, reappears. And doesn't just reappear as itself, but is rejuvenated, in a form that has been so useful in so many ways to so many peoples that in a certain sense, willows and people evolved together. People working willows, from sleeping mats to fishing creels, engage in an endeavor requiring close observation of the response of the species to the removal of stems and leaves, and the utilization of this compensatory growth to manufacture an enormous number of useful and essential items. The production of long flexible shoots, called rods or whips, is an old technique, an ancient gift from the vegetable world. This characteristic has been recognized and employed by peoples all over the world, including California.

BASKETRY CULTURE

To deepen your understanding of willows, go to the basket weavers. To understand coppicing (cutting back to the ground) and pollarding (cutting new growth back to the trunk), follow the basket makers. In California, other parts of the country, and throughout the world, willows were and are assiduously coppiced and burned to produce the long flexible shoots required for the foundations and threads of both twined and coiled baskets. In England, coppicing is used to produce long flexible rods for basketry, while pollarding is used to produce long rods for fencing. The culture of these osier beds is another arena in which the English display their considerable horticultural skills. A single willow crown can produce an annual crop of rods for thirty or forty years, though manuring is recommended for such intensive use. A painting by Vincent Van Gogh, *Landscape with Pollard Willows* (1884), documents the French use of pollarding.

Some species are more useful for such practical purposes than others. In California, different tribes traditionally favored different *Salix* species: sandbar willow *(Salix sessilifolia)* was used by the Pomo and Coast Miwok, *Salix lasiandra* by the Panamint, red willow *(Salix laevigata)* by the Yurok, Karuk, and Hoopa, and Geyer willow *(Salix geyeriana)* by the Modoc.

Native Californians also used willow to make dwellings, acorn granaries, fish stands, racks for drying and cooking food, gambling and accounting sticks, light hunting bows, baby carriers, fire hearths and drills, toys, clothing, cordage, cradle board padding, ceremonial medicines, and, of course, the all-important pain-relieving salicins made from leaves or bark to cure colds, headaches, laryngitis, fevers, sore throats, diarrhea, skin problems, stomach problems, and rashes caused by poison oak. A not insignificant number of uses for a single genus.

WILLOWS IN THE RESTORATION GARDEN

The adjective "willowy" is high praise; to look like a willow is a good thing, signifying flexibility, slenderness, and grace. Who wouldn't want willows around, with their connotations of abundant moisture, rebirth, strength, and flexibility? Except for weeping willows *(Salix babylonica),* this genus has long been excluded from consideration in the garden, thought to be too large, too dependent on water, too uncontrollable. Fast-growing and indestructible, shapely and gorgeous, our native willows can be a valuable addition to the larger garden.

Place-names with the word "willow" in them are appealing, almost as frequently used in developments as "oak." Associated as they are with cool, watery places, in leaf willows are everything that is refreshing. In the fall, with the leaves gone, the bare stems reveal buds of many hues, from the tenderest salmon pink to rich reds and sumptuous yellows. The bright red galls that hang from willow leaves are sometimes numerous enough and noticeable enough to be a positive visual attribute.

From a distance, a grove of unleaved willows is a landscape feature to be reckoned with. A contrasting island of creek dogwood, with its mahogany stems, makes for a heart-lifting large-scale planting, reaching its showiest moments in the fall and winter. Many species don't actually need streams or wetlands to thrive. Though they like wet conditions, they do not require them. They may flourish on dry hillsides and in places where no obvious sources of water are available, discovering groundwater where it was not apparent. In this, they are like my easygoing ancient cat, who likes being petted but does not require it. This important attribute of some plants associated with watercourses should be noted by the gardener.

Upright shrubs or multistemmed small trees, there's a willow for everybody. For the smaller garden, there are smaller willows, including alpine, desert, and coastal species.

Growing your local willows is another way to fine-tune your residence in your place, along with growing clarkias, poppies, oaks, and grasses particular to your area. Willows are tricky taxonomically, frequently referred to as a "taxonomic nightmare" because they intergrade from one species to another, with plenty of hybridizing to confuse the issue. Even if you aren't sure which species is which, what matters is that you take cuttings for your garden from the closest appropriate willow grove. Note that willows are dioecious, so some plants will be male, some female. Male plants are reputed to be better for basketry. If you have room, take cuttings from both male and female plants, in consideration of the theory that allergies are the result of too many male plants with no place for their pollen to go.

PROPAGATION AND CULTURE OF WILLOWS

Once the long, narrow leaves have fallen, it is time to propagate new plants. It has taken me years to comprehend just how large willow cuttings can be and still thrive. In fact, in my experience, large, well-branched cuttings outperform whips with no side branches. We put them in five-gallon cans with loose potting soil and by spring, when they leaf out, their roots have begun to fill the container. If I'm putting cuttings directly into the ground, I wait till the leaves are gone, cut branches up to two inches in diameter, and jam these branches two to six feet into the ground, where they turn into instant shrubs in a couple of years or even one year, depending on aftercare.

People love to tell stories about the legendary rooting abilities of willow cuttings. One restorationist working on the Mattole River pushes nine-foot-long cuttings six feet into the river bank, so that only three feet are exposed aboveground. They weather winter storms and floods, remaining stable enough to allow rooting to take place that will hold the riverbank. Smaller whips, he said, don't root as well. Some propagators use a solution, called "willow water," made by soaking willow twigs in water, to encourage cuttings to form roots. Landscape architect Nancy Hardesty made wattles out of woven willow whips to heal gullies in the forward-thinking development called Portola Valley Ranch in Woodside, California. "Bioengineering" with willow mattresses is a technique being used to restore damaged creeks.

Willows also have value as windbreaks and hedges. Newsholme recommends staggered pruning, where alternate trees are pruned to the ground during successive years. "Windbreaks treated in this way withstand the most powerful gales, and, although deciduous, their twigs provide a very effective wind filter and shelter belt in the winter months." Willow roots will seek moisture and should be planted well away from pipes that carry water, drains, septic systems, and irrigation pipes. If the garden isn't large enough to allow a distance of fifteen feet between a willow's ultimate drip line and pipes that carry water, then eschew planting willows except in containers.

WILLOW SOLUTIONS

I live in a town of ditches. A high water table combined with individual, frequently problematic septic systems seems to require moving water off the land. I want to move water off my land only if absolutely necessary; so far, I have resisted draining my land with ditches. I want my land, and all the land around here, to absorb as much moisture as possible, while cleansing the water and the land of contaminants. The willows can help, along with the other wetland plants, the numerous and useful evergreen sedges, which are also indispensable basketry materials, as well as creek dogwood, twinberry, and red elderberry. Wetland plants are also used in this way by the town of Arcata in Humboldt County in its system of sewer ponds.

Continuing to drive south, and continuing to muse about willows, I stop and investigate willow copses along the way. In some of the places where willows should grow, eucalyptus has replaced them. Eucalyptus can take the place of dryland and wetland species alike and, like willow, imbibes massive amounts of water. Both species suck up moisture, but willows have proven that they can coexist with creeks both seasonal and year-round, *using but not using up* their moisture reserves.

I crawl into these willow fastnesses and to my surprise, though we have not yet had rain, I hear Pacific tree frogs, who also like the conditions that willows both seek and foster. These frog sounds have diminished in my neighborhood of late, and it reassures me to hear them. Shading and holding and influencing the chemistry of water, can willows help heal what's wrong with the frogs? Might willows contain frog medicine?

As I continue this willow-rich drive, these copses invite me to crawl into their tangled limbs for rest and healing. I smell the sweetness of the cambium layer, which is reputed to contain the highest concentration of salicin. This ingredient of which aspirin is made is also an important compound in producing fragrances. The smell of the inner layer of willow bark evokes for me a strong childhood memory. At age four or five, sitting at the edge of the drainage ditch that wound through our midwestern neighborhood, I am industriously peeling willow twigs and savoring the sweet smell that is released by this activity. Kids, like monkeys, are natural peelers. Hearing my mother call me for lunch, I reluctantly leave this activity.

In another culture, I might have brought my mother these peeled twigs, to be further split and stored in coils for future basket-making activities. In this culture, I go in to have lunch, bringing the memory of my first tactile experience with a plant. All over the planet, in our giant planetary watershed, different species of willows grow in the world's watery places. They hold the shape of watercourses of all sizes and kinds, provide homes and food for insects, amphibians, birds, and ungulates, are factories for materials with which humans make containers, shelters, and toys. They sing a beautiful song with their rustling, easing our eyes with their foliage and our joints with their chemistry. Given this long association, our genes may make us crave willows; if you have the room, your garden can be a place to continue our long mutual residence on this planet.

7 The Landscaping Ideas of Jays

Every bird going upslope bore an acorn lengthwise in its bill; every bird in return course was empty-billed.

Joseph Grinnell, "Uphill Planters," 1936

The aim of restoration *is* noncreative. . . . Once past that, the value of the deliberately *non*creative act as a stilling of the will, an expression of obedience and humility, and the entrainment of consciousness to the gesture and movement of the other—an important element in religious practice—becomes clear.

William R. Jordan III, *The Sunflower Forest,* 2003

In 1934 eminent naturalist Joseph Grinnell sat by the lower Kaweah River in southern Tulare County, watching bright blue western scrub-jays zoom uphill. Each beak held a single acorn, plucked from the bounteous blue oaks along the riverbank, to be cached in the dry upland hills. All day long, jays repeated this journey, flying uphill with an acorn, returning, as Grinnell described it, "empty-billed."

Previously, climbing the western slope of the southern Sierra Nevada, Grinnell had observed a bumper crop of black oak acorns dropping from the trees, tumbling down the mountain, round, smooth, and heavy. "In that place and on that day I saw no acorn moving uphill. Gravity alone was acting as the agency of distribution." Without animal help, such as that provided by the western scrub-jay, acorn woodpecker, chipmunk, gray squirrel, or ground squirrel, our upland forests, woodlands, savannas, and scrub-oak chaparral would not exist as they do today or, when burned in the inevitable fires, would not be replaced. On that day by the river, Grinnell understood something that had been unclear to him until then: the role of the western scrub-jay in the landscaping of California.

Grinnell's reflections, which may seem obvious to us now, were not commonly understood in his day. Only thirteen years before, another well-known ornithologist, William Dawson, in his classic *Birds of California,* called this critical ecological role a minor function of the jay. "Doubtless," he says, "this miserly trick (of burying acorns) has served nature's purpose now and then in producing new trees." To the native peoples of California, the importance of the jay's role has been apparent for a longer time. According to C. Hart Merriam, "The Middle Mewuk of [the] Stanislaus River region say: Ki'-ki'-ah, the Crested Bluejay of the mountains [Steller's jay], plants acorns so that oak trees come up almost everywhere. (Several other tribes mention the same habit which, by the way, is hardly a myth)."

In my own West Marin garden, I've enjoyed the results of scrub-jay seed planting for twenty-five years. Of the nine coast live oaks now growing here, I take credit for the planting of only three; *Aphelocoma californica,* the western scrub-jay, accounts for the rest. California hazel as well turns up far from the parent shrub, frequently in the hospitable embrace of coyote bush. Coffeeberry, California bay laurel, and California black walnut are other species I find growing here and there. Like acorns, their seeds are too large to be moved by ants and too dense to be carried by the wind. Just a short way inland from here, the acorn woodpecker might be responsible, or the chipmunk, gray squirrel, and ground squirrel. Other animals engaging in caching behavior in California include chickarees, wood rats, Steller's jays, and band-tailed pigeons. Here, on this marine terrace, scrub-jays—which range all over the West south of Oregon—are the primary tree planters, acting out of an implacable desire to bury things, paired with the intermittent ability and occasional necessity to retrieve them.

The trees and shrubs that grow from the seeds that the western scrub-jay caches in my garden are generally allowed to stay in place. I like to think that in so doing, I am honoring the landscaping ideas of jays. This device is a way of thinking about a kind of gardening that helps us become aware of the actions, needs, and impact of butterflies, deer, ants, bees, jays, and numerous other wild creatures, who become, in a sense, our gardening partners.

Western scrub-jays need to be good but not too good at acorn retrieval. The more acorns and nuts they cache and the fewer they retrieve, the better for the spread of oaks, hazelnuts, walnuts, and various pine trees. Yet they must be able to retrieve enough to maintain viable populations, to feed themselves during lean times. The mechanisms of retrieval of stored food for western scrub-jays, as well as for several other species of corvids, have been the subject of many ornithological studies since Dawson and Grinnell pondered these matters.

Ornithologists credit scrub-jays with only moderate success in acorn retrieval. They are not geniuses at it, like their corvid cousins, the Clark's nutcracker and the pinyon jay. Clark's nutcracker *(Nucifraga columbiana),* as described in Ronald Lanner's book *Made for Each Other,* relies heavily though not totally on the nuts produced by the whitebark pine *(Pinus albicaulis),* which returns the compliment by being utterly dependent on Clark's nutcracker to open its tough and unyielding cones and bury the seed in appropriate places. One bird can cache as many as 98,000 seeds in over 30,000 caches in a year. The nutcracker has dazzled researchers with its uncanny memory, locating specific cracks and crevices where nuts were stored up to ten months previously. Reputed to move up to 55,000 of these wingless, gravity-bound seeds per month, the bird recovers and eats only half of them. The rest have a good chance of germinating when summer melts the snow and raises temperatures. Says Lanner, "Working in concert, Clark's nutcracker and the white-barked pine build ecosystems."

Researcher Russell P. Balda, who has studied extensively the caching abilities of corvids, says he once tested a graduate student on the retrieval of cached acorns. "He didn't do half as well as the nutcrackers did," recalled Balda.

It should be noted that the birds had considerably more "reason to know" than the graduate student, to whom the hidden acorns were of only academic interest. If the search object had been the keys to a Mercedes, or a contract for a tenured position at Harvard, the graduate student might have shone, the camp robbers putting in a lackluster performance. To a nutcracker, each shining acorn is the equivalent of the Nobel Prize in ornithology. The pride he feels in each accurate retrieval equals relief from hunger and the password to living another day, to squawk at a dog, chase cats, or sneak into a covey of quail to pretend he is one of them for a while.

Western scrub-jays don't need to be as good as Clark's nutcracker, which lives in more extreme climates. The Bay Area and other similarly moderate climates provide the western scrub-jays with many other things they like to eat throughout the year, including insects, lizards, apples and pears, seeds of all kinds, wax worms (a particular favorite), cookies, bay nuts, bird feed, spiders, earwigs, cicadas, snakes and salamanders, elderberries, gooseberries, dog food, cat food, cattle feed, and, yes, the eggs and babies of other birds. The western scrub-jays' diet also includes ticks, which are reportedly removed directly from the backs of mule deer and black-tailed deer, who "facilitate activity by standing still and erect, with ears raised," according to *The Birds of North America.*

Western scrub-jays, like Clark's nutcrackers, use landmarks, such as trees and rocks, to find cached food, and the sun as a compass by which to orient themselves. Experiments have shown that scrub-jays have memory of where food is stored, when it was stored, which stored items have been retrieved, and what kind of food item is stored. Individual scrub-jays who have pilfered the caches of other scrub-jays even demonstrate awareness that they themselves are being watched while hiding acorns. Waiting till the observer goes away, they then re-hide the nuts elsewhere. All of which implies the sophisticated ability to understand, "What I have done unto you (rifling your cache), so might you do unto me." Innocent scrub-jays who have not engaged in pilfering behavior do not as a rule re-cache.

THE JAY'S DUALISTIC NATURE

Though the western scrub-jay is significantly responsible for our oak- and pine-studded hillsides, the bird does not arouse only gratitude. In 1923 William Dawson had quite a bit to say about the jay's dualistic nature. "He, the malaprop, the impertinent, the sly wag, thief, scoundrel, outcast, jackal of the bush . . . as innocent as morning, as industrious as noon, as wicked as night." Dawson calculated the number of bird eggs a jay may be responsible for eating and wondered that nature can absorb such predations. Because of the jay's interest in the eggs and young of other birds, Dawson claimed that the jay is "undoubtedly the chief biological control factor in the distribution of bird life throughout the area specified." Tom Gardali of PRBO Conservation Science (formerly the Point Reyes Bird Observatory) in West Marin has a more up-to-date assessment of this aspect of the jay's reputation. Says Gardali, "This is likely not so; most nest camera studies show that corvids are relatively minor nest predators—snakes seem to be the stars." Still, jays are stunningly observant and, according to Gardali, can learn to spot and follow an ornithologist on the way to study a bird nest, in anticipation of a tasty bird egg or baby bird treats. Researchers at PRBO have learned to head off to study sites only after making certain they are not being spied on by a scrub-jay. In fact, contracts between ornithologists and funding agencies require that no western scrub-jays or cowbirds be in the vicinity when field work is performed.

Geoffrey Geupel, director of the Terrestrial Ecology Division of PRBO, contends that it is in the venue of the bird feeder that jays exhibit so much of the behavior that bird watchers dislike. (In general, Geupel is not a fan of the bird feeder. "Use the money you would otherwise spend on birdseed and bird feeders to restore habitat," he says.) I recommended to several clients who complained to me about jay behavior that they eliminate their feeders and focus instead on appropriate native plantings; it seems to work. Using bird feeders with openings and perches too small for scrub-jay use is also effective. Open compost piles are another jay attractant that can encourage out-of-balance populations of jays, as well as crows and ravens. Gardali says, "Jays and other corvids only really have the potential to be a problem when their populations are subsidized to the extent that they grow beyond normal levels—subsidies include a solid food supply (e.g., feeders) during times when food would naturally be scarce (winter). Or more food during the breeding season could lead to large clutch size and double brooding."

A STEADFAST DISLIKE OF JAYS

In a backyard study of the habits of jays written in 1945, the authors Harold and Josephine Michener describe in detail the caching behavior of the western scrub-jays in their garden. "The whole effort seems to be directed definitely toward getting the object out of sight," and then placing a clod of dirt, a rock, or leaf over it. Later in their research, annoyed by the noisy and domineering behavior of the male, this backyard husband-and-wife research team traps and kills him, after which, they say, "Everything seemed peaceful among the birds in the yard." Unfortunately, the surviving parent is unable to both feed and protect her offspring, who are soon found dead just outside the nest.

Jay hunts organized in the 1930s, for the ostensible purpose of retaining high quail populations, are another example of historic scrub-jay prejudice. Yet I frequently find scrub-jays in the middle of quail flocks, peaceably foraging for seeds and insects. Native peoples knew the scrub-jay as a thief but also as a sentinel, who let no intruder appear without warning. When the quail covey is disturbed and scatters, the jay flies to a nearby tree, screeching or screlching indignantly on behalf of his friends. When I put acorns out for the quail, they leave them untouched until the jays come along. Using horizontal branches as anvils, the jays are able to crack the nuts open, and the otherwise inaccessible nutmeat that they inadvertently drop or leave behind only then becomes available to the quail.

In my garden, I will always remember which trees were transplanted from containers into the ground and which had the luck to grow from acorns planted directly in the ground by the beak of a cheeky corvid, whose apparent intent was to return for a later snack. I honor their planting choices partly because trees planted directly in the ground by jays have a different and usually healthier trajectory through time than the ones that begin their lives in nursery containers, to be subsequently transplanted into the ground with some degree of care and coddling. Trees that grow from acorns planted directly in the ground, pounded in by the jay's sturdy beak, their roots never for a single moment confined by a container, have a significant advantage. Once, that is, they make it past the seedling stage. Acorns in the ground are vulnerable to many kinds of hazards, from rot and mold to weevil attack and, of course, retrieval by squirrels and other rodents as well as the jays themselves.

Relinquishing a bit of design control to these perspicacious birds may be appropriate for situations other than gardens. In many restoration projects, acorns are harvested, hauled from collecting grounds, cleaned, refrigerated, taken to the planting field, sown in the ground, and protected with screens and tubes from deer above and gophers below. Perhaps time and effort might be saved if the acorns were left in a huge pile near the area to be restored. It might not be long until, like an Andy Goldsworthy sculpture of twigs and stones borne out by the tide, a river of birds pours down from the sky to deconstruct the pile and find for each acorn a likely resting place. One bird- and oak-lover in Modesto, living in what was once the floodplain of the Tuolumne River, put out hundreds of valley oak acorns for the local scrub-jays. In the course of three days, dozens were removed and cached all around her house, garden, and neighborhood, in loose soil, leafy debris, roof tiles, gutters, and planter boxes. Across from her house, a lone oak, the source of this bounty, is silhouetted against the western sky, the last survivor of what was once a dense riparian jungle. Given a chance, the western scrub-jay could return the original vegetation, shade the houses that now bake in the summer heat, and re-landscape the neighborhood.

Such a project would also provide an opportunity for what the Cornell Lab of Ornithology calls "citizen science." Not all backyard studies require murder. With their tolerance for our nearness, jays make good subjects, which I experienced myself one day while cleaning seeds in our seed room with the door open. Spotting a bowl of coast live oak acorns, a jay swooped into the dim cool room, grabbed an acorn, and left with it. Thirty consecutive trips to the seed bowl followed, as thirty acorns were removed, one by one, from the bowl. (Unlike Clark's nutcracker, whose sublingual pouch can accommodate more than eighty whitebark pine seeds, the scrub-jay can carry only one acorn at a time.) My presence, three feet away from the bowl, did not deter him.

Covering the bowl with a cloth, I waited for the jay's return, curious to see if he would employ complex retrieval behavior in my presence. He assessed the situation, then left, returning three minutes later to see if things had improved. When I pulled back a bit of the cloth so that some of the acorns were in view, he grabbed one of the visible nuts and left. Covering the bowl again, I waited to see if he would pull the cloth back himself on his return, but apparently this activity required too much time in my presence. Even for the cocky jay, this proximity pushed the comfort zone.

Our unplanned interaction, more play than experiment, lasted for almost an hour, with about two minutes between raids, presumably for caching nearby. Which factors brought the jay to investigate the seedroom that day can only be guessed at. A silent watcher, he may have picked up on pertinent details, such as my journey from car to office carrying bags, that proved to be profitable information for him.

FORESTS 'R US

In my weakness for jay-planted oaks, in the early years of my garden, I often let them stay where they were, though they have since turned a greenhouse into a shade house and kept a new addition small. Sometimes, I find advantages later that were not at first apparent. One jay-planted oak nicely frames my garden; another provides a nighttime roosting site for quail so that their nocturnal cooings and rustlings can be heard from a bedroom window. I didn't anticipate these pleasant consequences. In this way, my design decisions are aided and abetted by the creatures that share this territory.

Perched at the top of one of the trees planted by an ancestor, weeping, screlching, zeeping, zraanhing, rattling, chukking, or shlenking, the jay encourages me to continue to consider his landscaping ideas. I regard each small seedling, whether an oak, California hazel, bay, or coffeeberry, as something more than itself, a proof of continuance of a partnership, not at all trivial, still at work.

8 In Praise of the Unleaving

Deciduous Trees and Shrubs for the California Gardener

MÁRGARÉT, áre you gríeving
Over Goldengrove unleaving?
Leáves, líke the things of man, you
With your fresh thoughts care for, can you?
Gerard Manley Hopkins, "Spring and Fall," 1876–1889

How beautiful the leaves grow old. How full of light
and color are their last days.
John Burroughs, 1913

Though I live in California now, I know the gold standard for fall color, for I once resided in New England. In that cold clime, the sound of the geese flying south in November plucked out my heart and returned it slightly different, partly in fearful anticipation of winter, and partly with exhilaration at the intensity of the northern seasons. Immersion in the visual flood of a huge sugar maple in full florid hue, beech trees golden in the woods behind my house, the red oak, the tamarack on the hill, and the "leaf-peeper" traffic jams, all these I remember.

In California, leaf viewing is a less well-recognized phenomenon. We have wildflower hotlines in the spring, but no leaf-viewer hotlines in the fall. (East of the Sierra, still within California's borders, is the exception. Aspen viewing in the fall is commonly acknowledged to be spectacular.) The counties with milder winters provide milder experiences of leaf viewing. Still, I find in the more modest unleavings of California's Floristic Province, in her less well-sung list of deciduous species, ample reason to viscerally declare a new season. This list includes trees and shrubs that may, in a good year, astonish, but also a longer list of those with gentler changes. Slower, subtler color events are characteristic of fall in California, including a rich array of beiges and browns and many shades of yellow, from bright gold to tawny honey and mellow lemon. I relax into these calm yet luscious hues and into California's flowing seasonal changes, knowing I am in good hands. I won't be bored, and I also won't freeze to death.

THE SCHEME OF THINGS

From the western redbud's hues in the mountains and foothills to the rattling beige leaves of California sycamores along Central Valley streams, many trees and shrubs that lose their leaves in the fall in a more or less notable fashion are available to the native plant gardener in all parts of California. Besides western redbud *(Cercis occidentalis),* those species undeniably showy in fall include vine maple *(Acer circinatum),* bigleaf maple *(Acer macrophyllum),* box elder *(Acer negundo),* valley oak *(Quercus lobata),* black oak *(Quercus kelloggii),* and certain selections of California wild grape, most notably *Vitis californica* 'Roger's Red'.

Yet the undisputed star of our fall season is poison oak *(Rhus toxicodendron).* Growing most places in California except the desert, not choosy as to soil, exposure, or plant community, poison oak is the first native plant every California child must learn. Though dedicated is the gardener who would actually plant poison oak (and it has been known to happen), some are brave enough to leave it in place where it already exists. In the wilder regions of a large garden, its fall hues, ranging from brilliant scarlet to deep, soft maroon, may be enjoyed from a distance. Metal barriers can be installed belowground to limit the expansion of poison oak thickets.

The removal of poison oak where it grows in proximity to human activity is understandable, yet poison oak berries are a critical winter food source for many birds. The pale berries, though not showy to us, are of great interest to woodpeckers, jays, and especially the wrentit, that home-loving bird whom Dawson calls "California's native son." Ninety-two percent of these nonmigratory birds live and breed within California's borders. Their range closely corresponds with the range of poison oak, a species that also might justifiably be called another of California's native offspring. Wrentits no longer breed in San Francisco's South Bay, the cessation of this phenomenon coinciding with the extensive removal of poison oak.

In California's colder counties, deer depend on poison oak for browsing, and their close cousin, the domestic goat, prefers it above all else. A client in the Sierra foothills rolls large balls of poison oak clippings down the hill and directly into his goats' corral, where it is devoured with relish. As many gardeners know, after being cut, poison oak resprouts reliably. Except for considerations of toxicity for the goatherd, a reliable source of browse is permanently available for goats with poison oak on the property.

Poison oak is a hard sell in a small garden, particularly to the sensitive, which includes most of us, but the exurban gardener dedicated to maintaining and enriching the scheme of things might consider leaving it in place. Though hardly an endangered species, it plays a vital role, and the gardener who includes it in the garden is automatically given membership in an elite club.

Part of being a Californian is learning the poison oak dance, sidling around it, under it, and through it, while managing to never make contact. It seems impossible at first, but it can be done and frequently is. One friend, who lives at the base of a coastal mountain, consciously welcomes what she calls "the three poisons" into her garden. Poison oak is one, the western rattlesnake another, and the third is ticks. According to Jaime d'Angulo, among the Achumawi of Modoc and Shasta Counties, the word for power is the same as the word for poison and also spirit helper, pet, and medicine. All three problematic species confer a kind of power upon those who can coexist with them. The willingness to live with "poisons" indicates that a person of substance lives here.

THE SUBTLER COLORATIONS OF CALIFORNIA'S FALL

The demarcation between seasons in California is not sharp. Seasons are long and leisurely, with much intermingling of characteristics. The eye is drawn into the depths of shrubs not famous for their fall color that yet, with their subtle changes, provide a marked pleasantness to the season, with details that are well worth noting.

Some deciduous species have leaves that color from the tip to the base as they lose chlorophyll, while still others send random blotches into the leaf. Much depends upon whether color comes from the withdrawal of chlorophyll or from some combination of rust and decay, or both. All these have their beauty.

Alder leaves, for example, initiate their drop by coloring yellow at the base where the stem attaches. As the season progresses, the color flows to the tip, till all is yellow. Willows, their partners at the riparian edges, hardly register the change of season till long after alder leaves have colored and dropped. They turn pale gold with yet still some green in it, a blending that epitomizes the way California's fall color, much like her seasons, is a subtle combining. Masses of willows are a design element that can be used effectively in the larger landscape, along with creek dogwood. Once the leaves of these species drop, the new growth and buds become visible, displaying showy hues of salmon pink, deep red, or yellow that light up the fall and winter scene. Since both are thicket-forming, planting a few will eventually give many.

Red elderberry leaves blacken gradually, beginning at the tip of the leaf and proceeding to the base. Pink flowering currant holds on to its leaves for a while, as their edges crisp with rich rusty brown, here and there a patch of true ruby. This edging of terra-cotta is provided by the disease called rust, whose colors I have learned to appreciate. Almost as early as the currant, thimbleberry creates a lush leaf litter with pale ocher-brown leaves that first lighten almost to chartreuse before dropping to reveal creamy beige twigs. Vine maple is famous for its sharp colors in northern California and sometimes in the south as well. The strongly veined leaves of creek dogwood turn, in colder years, a rich burgundy, well worth bringing into the house; in other, less frosty, years, they are a more somber yellow brown. California hazel, California black walnut, wild grapes, poplar, cottonwood, alder, all add their vivid or somber note to fall's particular statement.

The forbs—broad-leaved annual plants—also have their moment. While driving through the woods in Samuel P. Taylor State Park (Marin County), I am stunned by a blaze of yellow, which upon examination turns out to be the yellowing fronds of bracken fern. In some places, the blades of Douglas iris remain mostly green through the season, but in others, the blades turn cinnamon brown, a cinnamon matched by the old fronds of nearby western sword fern that have not been removed in a zealous cleanup. If green alone is preferred in your garden, cut them to the ground in October or November and new green leaf blades will quickly emerge. Otherwise, they can be left to add their note to the season's symphony of muted hues, which, though quiet, is nonetheless delectable.

The gray brown haze of the annual elegant madia, with its finely articulated pedicels and twigs, can last well into the rainy season. In garden soil, it may grow six feet tall. Other annual wildflowers, those that haven't shattered with the summer heat, are still with us. In particular, the curved rust and beige seed capsules of clarkias punctuate a meadow in the fall. Notable among the clarkias for interesting seed-capsule arrangements that hang on long into the winter is showy punch-bowl godetia *(Clarkia bottae).*

Among the vines, let us speak first of the colors of wild grapes. Most remarkable is the soft burgundy of *Vitis californica* 'Roger's Red', which is thought to be a hybrid with the domestic grape, and the mellow gold of the straight species. California pipevine *(Aristolochia californica)* also adds a soft golden note against a weathered gray fence. On the bare branches of the blue elderberry, western virgin's bower traces the shape of the limbs with its fuzzy seed balls, in which dewdrops linger well into the morning. Growing on a trellis, this deciduous vine begins to lose its leaves while the seeds are still sparkling on the stems.

THE BEAUTY AND FUNCTION OF LEAF LITTER

Evaluating plants in terms of fall color must include not only the leaves on the tree but also those that have already dropped. Back East, leaf litter may soon be covered by snow, but we Californians frequently get to track more of the whole slow, leisurely progress from leaf to duff. Leaf litter is a very valuable substance indeed. *Litter,* with its connotations of that which is undesirable, synonymous with waste, mess, refuse, trash, disorder, requires a complete redefinition. For the gardener, it can provide pools of subtle coloration reminiscent of the colors still on the trees, forming shapes sometimes exquisitely matched to the drip lines of the trees from which it fell. As material, it is priceless, useful for defining areas, including paths, borders, beds, and islands, and irreplaceable as a component of soil health.

In the well-set-up garden, leaf litter can function as a gift to be treasured and hoarded. There should seldom be a reason to ship it off to the dump. Returning to the soil what the plants need, it also provides the labor-saving benefits of fertilization, conservation of moisture, equalization of soil temperature and moisture, and weed suppression.

This gift of leaf litter might become an annoyance, requiring continual removal, should it drop on hardscaping, that is, on human-made structures like cement, brick or stone sidewalks, redwood decks, lawn chairs, and glass tables. Limiting the area that requires sweeping and expanding the amount of space that looks good with leaf litter is a helpful strategy. Note that evergreen trees and shrubs continually drop leaves in small amounts, while deciduous plants will "unleave," with some exceptions, during a month-long period.

Lawns are the other situation in which leaf litter requires removal. The case against the conventional lawn has been made frequently in the literature and will not be repeated here. Its incompatibility with leaf litter is only one of many possible indictments.

The coast live oak frequently puts out new growth in the fall, soft pale leaves with no hint of the defiant prickliness they will later develop. At the same time, it continues its slow, regular shedding of spiny leaves. As one always on the look-out for sources of mulch, I can only be grateful when the coast live oaks deliver their weed-suppressing, unifying, texturally pleasing khaki and beige leaf litter right to my door. We wade through wonderful masses of it, though usually not barefoot. Long lasting, it is reputed to require the work of the oak leaf moth caterpillar to break it down, which, coming every seven years or so, finally releases nutrients to the soil.

Another of my favorite "litterers" is the genus *Alnus,* which includes redwood alder *(A. oregana)* on the coast, white alder *(A. rhombifolia)* inland, and mountain alder *(A. tenuifolia)* in the mountains. Alder's leaf drop is among the showiest and most dramatic in the garden. Its madrone-colored leaf litter draws the eye, even while many of the leaves remaining on the tree are still bright green. This pool on the ground creates a sharply defined mirror of the tree's outline, soon to be processed by the winter rains, churning the small bits into a gooey muck.

Alder leaves are a good example of the supreme value of leaf litter. Studies show that one acre of alder trees adds from one hundred to five hundred pounds of nitrogen to the soil through the leaves they drop in the fall. The remarkable nitrogen-fixing capacity of their roots, especially during the first fifteen years of their life, adds to their reputation as "soil doctors." Quick-growing, fast-rooting alders can be medicine for clear-cut, eroded land, says Ray Raphael of Humboldt County.

> As an added benefit, the roots also host certain mycorrhizae that minimize the incidence of root diseases, not only for themselves, but also for neighboring trees of different species. The red alder, in short, is an ideal healer of damaged land; it simultaneously penetrates, and stabilizes, and enriches the soil, while providing a kind of inoculation against disease for other trees nearby.

When the felted luxuriance of California hazel leaves appear on the ground, the buds for next year's female flowers have already appeared, mixed with male catkins and leaves still green, the confusing Californian mingling of the seasons.

Similar in impact are thimbleberry leaves on the ground, also tan and beige, decaying on a similar time frame.

Again, deciduous trees and shrubs are not the only ones worth watching in the fall. Coffeeberry, though evergreen, drops a luxuriant and interesting leaf litter at this time of the year. Its pale, ocher-colored leaves curl lengthwise like prerolled cigarette papers. Pacific wax myrtle, in its old groves, makes a couch worth reclining on, cinnamon-colored leaves sticking around long enough to make beds up to two feet in thickness where these large shrubs have been allowed to reach maturity.

LEAF LITTER AND BIRDS

Something there is at this time of year that does not enjoy a swept sidewalk. Every evening in the fall I sweep the leaves and twigs off the cement sidewalk, and the next morning it is as though I had made no such effort. Unfailingly, dirt and debris have been returned to cover and soften the hard edges of this human-made substance.

Who's responsible? It's the thrashers and towhees, the sparrows and jays, those dedicated mess makers, churners of raw material, the upper gastrointestinal tract of the soil. Autumn's arboreal unleaving coincides with the brown thrasher's yearning for the insects, beetles, centipedes, worms, earwigs, and sow bugs that in their turn relish the crackly on top, matted underneath circumstances created by leaf litter. To get access to these goodies, birds must sort and resort debris, breaking it down into ever smaller pieces.

The colors of female ground-nesting birds take their cue from leaf litter. Coevolving with that which drops from the trees, ground-nesting birds frequently have feathers that match the browns, duns, rusts, and grays of leaves as they are transformed into soil. Writing about it, I can hardly resist going out to the garden to experience some leaf litter for myself.

In California we don't have to be quite as sad as the poet's young friend Margaret, or as fearful as I was at the onset of a Vermont winter. There's a luxurious subtlety to the shift, a slowness to the unleaving, not as sharp as back East, but gorgeously present. The senses are nudged to the change.

Winter

9 Eating the Rain

> This plant has been a favorite salad among the inhabitants of the Pacific Coast since prehistoric times, and from this use its various names of Indian lettuce, Spanish lettuce, and miner's lettuce may be readily understood. . . . In England, where it has been grown for over a century, the plant is known as winter purslane.
>
> Leslie L. Haskin, *Wild Flowers of the Pacific Coast,* 1934

Winter can be many things in California, every year different, but it is always a time of interesting skies, migrating cranes, long nights, and, hopefully, opportunity to slow down. Limiting activity in the winter, some sages say, allows us to replenish ourselves. Every winter an elder in my town used to put in our local paper a plea for a seasonal slowdown. The holiday season, she claimed, was a time for less, not more. The payoff for storing up energy in the winter would be an exuberant, explosive spring. If I lived in a clime that did not enforce a change in activities, I would consider enforcing it on myself, so much do I value these dreamy retreat times. Weather permitting, I weed as well, strengthened for this activity through "eating the rain," enjoying the green tonic of a winter native salad plant.

Here on the Pacific coast, seeds of miner's lettuce *(Claytonia perfoliata)* germinate soon after the first fall rains; by December the plants are ready to eat. The first leaves that appear are long, narrow, and delicate, followed by true leaves that widen at the tip to the shape of a spade. Plucked at this stage, they have a delicious freshness, a delicate and tender succulence, especially with dew or last night's rain beading the leaves. After a long time with no greens from the earth, they arrive as a welcome wake-up call: the great California germination of annuals begins again.

Found in moist shady places like creeks and ponds, around the drip line of oaks, in coastal sage scrub, northern coastal scrub, chaparral, and even in sunny wildflower fields with a high water table, *Claytonia perfoliata* is reputed to be high in vitamins A and C. During the gold rush, local Indian tribes, who were familiar with its use, reportedly fed it to miners with scurvy. Rather than commemorate that environmentally and culturally devastating period of California history by calling this succulent, tasty, and healthful little green "miner's lettuce," I prefer to use the common name of "Indian lettuce," but frequently find myself forgetting to do so. By either name, *Claytonia perfoliata* is distinctive for its perfoliate leaf, heart-shaped and hard to mistake. At flowering time, the leaf merges into a single plate-like unit, up through the center of which a raceme of tiny white flowers emerges. Indian lettuce is a welcome addition to any salad, a backpacker's delight, and the first wild food of many a California child.

Fifteen years ago I sowed a crop of Indian lettuce in a shady corner. Every year since then it reliably reappears and, with a little help, has tripled its area. We nibble on it December through March, then begin to collect its seeds. In spite of vigorous seed collection, we never seem to deplete the seed bank in the soil. Germinating in the same spot year after year, this annual can also move with surprising rapidity into other parts of the garden, so much so that the *Jepson Manual* refers to it as "rather INV" (invasive). Once we sell a packet of seeds to a customer, I'm fairly certain he or she won't need to buy another.

Indian lettuce makes an admirable salad green, requiring no planting, no weeding, no fertilizing, and no pest control. The problems that afflict my other salad crops, like snails, slugs, gophers, sparrows, not enough or too much rain, don't faze Indian lettuce. Some organic farmers are now including Indian lettuce in their salad mixes, and it has been sighted on the menus of certain upscale California cuisine restaurants. At the request of an imaginative local chef, we did an experimental harvest of Indian lettuce, learning that, to keep this succulent green plump with moisture, it helps to cut it just above the root so that the individual stalks are still connected.

Indian lettuce is good public relations for edible wild plants, reminding us that California's indigenous peoples stayed healthy on a diet of California native plants. The recent upsurge of interest in the protective phytochemicals provided by fruits and vegetables encourages us to take a closer look at the native diet, almost none of which has been analyzed for its potential contribution. Some of the easiest wild foods to incorporate are the berries, including huckleberries, elderberries, wild strawberries, salmonberries, and thimbleberries. The easiest green to incorporate is Indian lettuce, followed by creek monkeyflower, watercress, checkerbloom, and clover.

I control its spread by zealously harvesting its volunteers. I try not to let any occasion to serve Indian lettuce pass by. It can be mixed with spinach for a healthful salad, scattered around platters of hors d'oeuvres, or incorporated into appetizers, with its rounded shape set neatly on a round cracker topped with a round slice of cheese or a piece of local wild smoked salmon. A favored small hostess gift is a carefully harvested bouquet of tender Indian lettuce tied with a ribbon and accompanied by a small seed packet of the same.

One customer gave us the following recipe for Cream of Indian Lettuce Soup.

> In a saucepan, sauté baby onions in 1 tablespoon oil till sweet and limp. Add 2 cups of washed Indian lettuce leaves and 1/2 cup of sorrel leaves. Cover the saucepan and simmer briefly till the greens are wilted. Add 1 cup of homemade chicken or vegetable stock. Run through a blender or food processor.
>
> Add 1/2 cup of cream or milk. Salt and pepper to taste. For a heartier soup, add one boiled, pureed potato. Serve hot or cold, garnished with a sprig of yerba buena or coyote mint.

A client with serpentine soil, who was frustrated by her inability to grow salad greens in this lean medium, was able to find seed of Indian lettuce nearby, already adapted to her conditions. Indian lettuce was the solution to her dilemma.

THE GREEN TONIC

I begin each day for three months with this green tonic, and I am not the only one. Yesterday I watched from my study window as four different employees stopped at different times for snacks at the Indian lettuce bed. As much as we eat, there never seems to be less. Every year, the greens reappear in the same bed and also continually expand their domain. This year, potatoes were planted in the lettuce bed, and by the time the lettuce has gone to seed and been collected, the first young potatoes will be ready to dig. A garden bed with good soil will produce large-leaved plants with seed stalks 8 inches long, 31 to 36 capsules per stalk, 3 to 6 seeds per capsule, 100 to 200 seeds per plant. These crunchy black seeds are themselves edible, a treat for the pinole aficionado as well. When lightly roasted in a cast-iron pan for under a minute, they cook quickly, have a peppery taste, and hold unknown nutritional properties. After we harvest, along come the gleaners; first at them in the morning are the quail, to be followed by the mourning doves, and finally the sparrows in midday.

In his book *The Botany of Desire,* Michael Pollan presents the hypothesis that some plants (in particular, he looks at tulips, potatoes, apples, and marijuana) make people do things to further their vegetal reproduction. He ponders the question: "Did I choose to plant these potatoes, or did the potato make me do it?" It may be, as he hypothesizes, that many plants make many people do a variety of things, and Indian lettuce makes me collect its seeds. I cannot resist its intriguing way of producing and ripening seed.

First, there is the beauty of the little stalks (pedicels), which hold the individual seed capsules. This cunning arrangement is worth examining. They arrange themselves along the main stalk like vertebrae along a spine. Each pair emerges from the same point on the back of the seed stalk, but when you turn the seed stalk over to look directly at the capsules, so as to assess their ripeness, you see that the pedicels, though their origin is opposite, allow the seed capsules to nestle alternately rather than oppositely along the front.

Each capsule has two points, two little legs separated so that the point of the seed capsules beneath nestles between them. The capsules are roughly heart-shaped and carried toward the front of the stalk, so the back is knobby from the attachments of the pedicels, which head forward. The seeds ripen indeterminately, so that if you pluck a flower stalk with ripened seed at the bottom, there will be flowers or even unopened buds at the top. Before the seeds ripen, they are enclosed in a tough green skin, a round package at first bright green, then fading to yellow green. When the seeds eject, this skin remains behind as a tiny yellow brown desiccated container.

When this outer skin begins to lose color, the capsules can be made to split along two sides by applying gentle pressure. The skin opens in an impossibly precise, neat Y, releasing the shiny trio of seeds. This squeezing can be addictive, somewhat like popping the individual bubbles in bubble wrap.

Sometimes the skin that contains the seeds is bright green, but the seeds, when squeezed out, are already ripe and black. Before the capsules split open (or dehisce) along two seams, the capsules turn upward, so the seeds will rest for a time in the cup of the triangular capsule before falling soundlessly to the ground or, in some cases, to the plate formed by its own perfoliate leaves below.

The movement is this: the racemes bloom on upright stems, then as they ripen, they droop from the weight of the seed, then again right themselves to hold up the bright black gems till they fall. There is a faint dehiscing sound as the capsules split open, nowhere near as loud as lupines or poppies exploding, but a muted pop, which would hardly wake you up in the middle of the night, should they be drying in your room. If you do awaken, you might find it soothing, like a gentle rain rather than, as is the case with lupine seed, a gun going off. The margins of the capsule roll in as they dry, pushing the seeds out. A handful of these jetty black seeds glisten in my palm. The little round discs are so gleaming and shiny black that in the bottom of the seed envelope, they could be glinting chips of obsidian, except for the small white appendage at one end.

The *Jepson Manual* describes the reproductive parts of this species succinctly:

> Pedicels reflexed, becoming erect in fr. FL: petals 5, pink or white; stamens 5; epi-petalous; ovary chamber 1; placentas basal; style 1; stigmas 3. FR: capsule, valves 3, margins rolling inward and forcibly expelling seeds. SEEDS 3–6, gen black, gen clearly appendaged.

THE LANDSCAPING IDEAS OF ANTS

Whether or not *Claytonia perfoliata* makes me collect its seeds, it doesn't need me to disperse them. Some of the spread of this "rather INV" plant can be attributed to dehiscing seed pods, but the small white appendage that links the seed to the seed capsule indicates that another seed dispersal mechanism is at work. Known to botanists variously as an "aril," "caruncle," "strophiole," or, in the case of Indian lettuce, simply as an "appendage," this structure has a different name in ant specialist circles. Called an *elaiosome,* this structure appeals to certain ants, who use it both as a handle to make carrying easier and as a treat for their larvae. To those ants who seek it out, it's a delicacy, Gerber's baby food. Ants with this predilection are called *myrmecochorous,* and the phenomenon of which they are part is known as *myrmecochory,* the dispersal of seeds by ants attracted to the elaiosome.

The elaiosome comes in a wild variety, assuming as many different shapes, forms, and colors as there are galls in an oak woodland. It's a tasty part of the seed that the ants carry back to the nest (with the seed tagging along), a kind of "takeout" meal that provides dispersal service for the plant, as the seed can be dropped anywhere between the plant and the nest. In some cases, the elaiosome includes chemicals that resemble insect blood and stimulate ants to "picking up and carrying behavior." Like a bird finding a worm and bringing it back to its nest, the ant grasps the elaiosome in its mandibles and brings it, with the rest of the seed, back to the ant mounds, where the worker ants detach the elaiosome and feed it to the larvae. The seed, with everything it needs in order to germinate and grow remaining intact, is discarded in the fertile debris piles that ring the ant mound. The discarded seeds then have the opportunity to germinate in a well-aerated medium where they are relatively protected from predation from birds and rodents.

As well as distributing the seed of Indian lettuce, ants provided a related service for native peoples, according to Stephen Powers, in his *Tribes of California.*

> Of the wild lettuce a curious fact is noted. The Indians living in the mountains gather it and lay it in quantities near the nests of certain large red ants, which have the habit of building conical heaps over their holes. After the ants have circulated all through it, they take it up, and shake them off, and eat it with relish. They say the ants, in running over it, impart a sour taste to it, and make it as good as if it had vinegar on it.

You might want to know these things about Indian lettuce. In the middle of a California winter, it is the best way to eat rain.

10 Eternal Vigilance
Featuring Bermuda Buttercup

> It [*Oxalis pes-caprae*] is a rare and endangered plant in its home range of South Africa, where native fowl eat the foliage, a mammal digs the bulbs, and parasites infest bulbs.
>
> Jake Sigg, "Oxalis pes-caprae," 2003

If in a wet winter, storms persist, the water table rises, and gardening activities must be curtailed, I accept these restrictions gratefully and resolve not to notice spreading sheep sorrel or weedy grasses. There will come a day when all that will be addressed. I make an exception, however, for Bermuda buttercup, *Oxalis pes-caprae,* whose blooms I do not enjoy and with which I make an unrealistic but fervent refusal to coexist.

A friend calls to ask me to go to the cemetery with her. She wants to pick out a plot for her husband's ashes, a place where his favorite wildflower, pussy ears *(Calochortus tolmeii),* would be happy. We look for a sunny spot where no eucalyptus trees will drop smothering, tannin-filled leaves, where no cotoneaster will continually create hundreds of new seedlings. We need one where enough care is being bestowed on surrounding graves that *Oxalis pes-caprae* is being controlled, because the neglected graves are covered with it. Pussy ears could not coexist with this oxalis, also called Bermuda buttercup, yellow oxalis, sour grass, or cape sorrel.

As we explore the cemetery, which is tiny and unique, containing headstones from the earliest Spanish settlers to more recent New Age creations, a shift takes place in me. While talking to the grave-site salesman, an enthusiastic volunteer who has given countless hours to the restoration of the place, I begin to feel that it is no longer a matter of indifference to me where I am buried. It suddenly matters that I be put to rest among the plants that to me represent beauty, ongoing ecosystem health, and a long-term friendship.

How could I "rest" here if my grave was smothered and upturned by eucalyptus roots, as was happening till the townspeople, led by one valiant tree surgeon, initiated significant removal? Though his actions turned into another bloody eucalyptus battle, and though many of the trees were removed, enough remain to dominate the feel of the place. I note a reluctance at the thought of being buried next to those who fought for the eucalyptus. Though previously divided into Catholic, Presbyterian, and Methodist sections, perhaps the new cemetery will be divided along vegetative lines.

In my own garden, *Oxalis pes-caprae* is also a matter of concern. Walking my garden just before Christmas, I see that some of the areas I set aside for annual wildflowers are covered with tiny oxalis seedlings. This species, introduced as a garden plant, is currently enjoying a population explosion, one that most gardeners are not enjoying. Jacob Sigg, conservation chair for the Yerba Buena Chapter of the California Native Plant Society and federal issues coordinator for the statewide CNPS Invasive Exotics Committee, is a particular foe of *Oxalis pes-caprae.* As Jake said of his challenging restoration activities in San Francisco, "I didn't need oxalis." Its yellow blooms are now so prevalent that when customers come to me in winter asking the name of that beautiful yellow wildflower, I know without further investigation that it is oxalis to which they refer.

CAN GOPHERS FLY?

Much is mysterious about the spread of oxalis. These tiny plants can appear in masses that look like seedlings in their density and the amount of ground they cover, yet it is generally accepted that oxalis is not producing viable seed in California (though one case has been found). I am told that it does not even set viable seed in its homeland of South Africa, that only gophers can account for its devastating spread throughout the gardens, and now the wildlands, of California. Each plant produces many tiny bulblets, and it is easy to miss some of them when weeding. Even allowing for that reality, its spread is still baffling. It appears in wooden planter boxes with only tiny cracks in the bottom, in containers on tables, even in Christmas cacti kept inside a greenhouse, repotted in sterilized potting soil. It appears in rock formations that seem impossible for gophers to burrow through and has been reported in window boxes two stories up.

For years I have removed every oxalis plant in my garden in a timely fashion. Knowing that the older the plant is, the more opportunity there is for it to form little nutlets near the crown, and the more energy can be sent into the bulblets along the roots, the bulb at the root base, and the rhizomes, all of which can reproduce on their own, I have never rested on my laurels. Where oxalis is concerned, I have been vigilant. Says Jake Sigg, "Oxalis gives a new meaning to the word 'vigilant.' If you don't have oxalis in your garden, it just hasn't gotten there yet."

For years, I thought that oxalis was one of the easier pest plants in my garden to control. Though it appeared in large quantities every spring, I thought that if I continually removed either top growth or as much of the underground structures as possible, I could keep it under control. And for a number of years, it seemed that I was even making headway. Oxalis is a plant that allowed me to think I had controlled its spread. It is a plant that allowed me to think I understood it.

Still, says Sigg, "Removing a small infestation in a discrete area is feasible." Once an area seems free of it, he recommends monitoring for at least another two years, to make sure that no small leaves are helping an underground bulb to build up strength for a comeback. Perhaps we had better amend that time frame to something more closely resembling eternity.

On this pleasant early winter day, in the local cemetery with my friend, I say to her, "If I die before you, please don't let my grave be overrun by oxalis."

My friend gives no false assurances. "I'll do my best," she says, and with that realistic assessment I must be content.

11 Rain-time Reading

During the long nights and abbreviated days of winter, when rain may preclude work in the garden, we give ourselves rest. Maybe we have watched the salmon swim up the creeks to spawn and die, or witnessed the departure of Aleutian geese from the Sacramento Valley and the arrival of the gold-crowned sparrows along the coast. Short days mean less time to bustle about for the quail, who have not yet begun to couple up and sometimes seem to have disappeared altogether. Acorns have long since dropped and been cached or eaten, or are sprouting in the duff or in nursery cans, or are stored in refrigerators for later planting. Wildflower seed has begun to germinate, make sturdy roots, form good aboveground crowns, and then begin the wait for spring's enlivening sunshine. In between rain squalls, we may go out to see what's shaking, or venture out during warm rains from the south, which provide intermissions amenable to wildlife activity and viewing.

WINTER IS THE TIME FOR STORY

For those humans who can give in to the call, to settle in to roots already in place, to be content with darkness, winter can mean indoor work, short naps and short days, long sleeps and long nights, and good stories told to receptive audiences. For indigenous peoples in California, though hunting and other activities did and do continue in the winter, the tenor is lowered. Many of the native peoples of this continent mark wintertime as the exclusive domain of the storyteller. Various consequences, from snakebite to starvation to being struck by lightning, have been said to follow if stories are told at other times of the year. The Nez Perce tribe of Idaho told a child who asked for a story in the summertime that "a snake might come and 'visit' his home if he insisted on hearing the stories during the season when food was being gathered and stored for the winter." "If stories should be retold during the growing season," says an interpreter of Iroquois myth, William Spittall, "life must come to a halt as the friendly spirits of nature become enthralled by their magic spell and neglect their appointed function of providing sustenance for the coming winter."

Abenaki Indian Joseph Bruchac, in *Roots of Survival*, quotes Mohawk elder and storyteller Ray Fadden on the need for such proscriptions: a mother bird, flying by an Indian hut during the summertime, hears a story being told out of season. Becoming engrossed by the tale, she stops her food-gathering activities and hovers around the hut, listening, while back at the nest her neglected babies go unfed.

Things are a bit different in most of California, where winter, bringing the rain and weather that, in the California Floristic Province at least, infrequently plummets below freezing, also brings new growth. We must hurry to get in our stories before the rains bring up new shoots. Greg Sarris, Coast Miwok, tells us that Pomo doctor and basket weaver Mabel McKay, Long Valley Cache Creek Pomo, "mentions regularly that she cannot tell Coyote stories during the summer months. 'It is forbidden,' she says. 'It's an old-time rule. Us old people know that.'" Kashaya Pomo doctor and tribal leader Essie Parrish agreed. She and other elders "maintained that certain stories, especially those about Coyote, those from the time when the animals were still human, could be told only during winter."

In a colder part of California, where the Nim (North Fork Mono) live, Gaylen Lee tells of similar proscriptions: "My grandparents said stories were to be told only on the snowy days of winter so as not to disturb Rattlesnake, snug in his winter home, because he, in turn, disturbs all other living creatures. Rattlesnake would appear soon enough in spring, they said, when the orange-breasted flicker returned from his southern haunts and began singing to Rattlesnake that the days were warm, and he should come out and enjoy the sunshine."

EXPLORING THE CALIFORNIA NATIVE PLANT LINEAGE

In the winter, power failures, floods, debris flows, impassable roads, and other hazards make staying home appealing, with candles, stored food, dry firewood, and books close to hand. In the winter I read long hours, dipping into the California native plant literary canon. It seems that the three women whose lives and contributions I describe in this part, Lester Rowntree, Edith Van Allen

Murphey, and Gerda Isenberg, have been with me for a long time, inspiring and supporting my endeavors, and those of many of my fellow native plant lovers, though of the three I knew only Gerda.

The stories I am about to tell, like many traditional stories, are about our lineage, only in this case it's our California native-plant-loving heritage. Lester Rowntree, Edith Van Allen Murphey, and Gerda Isenberg had long productive lives that spanned between them 117 years and illuminated possibilities for other, younger, California horticulturists in the field. All three engaged in businesses having to do with native plants and all pursued their involvement, begun in middle life, well into old age.

Rowntree and Murphey were writers as well as seed and plant collectors. Isenberg wrote and lectured only occasionally but left a nursery that continues as a solid purveyor of native plants even ten years after her death. All three, outside any formal academic structure, promoted appreciation of California's flora when little attention was being paid by the general public. All three had personal histories that make good stories, worth retelling.

Lester and Edith were born in the nineteenth century; Gerda, the youngest, was born when the new century began and died ninety-six years later. Women ahead of their time, some common themes emerge from their lives: adventurous and unconventional, each came from stable, landed, to some degree moneyed backgrounds. (Rowntree writes: "My mother belonged to an age when 'gentility' meant a lot and was a word in better standing than at present.") They all were divorced at a time when divorce was not common. Two of them had a child with a serious physical ailment who preceded them in death. They all lived to be elders; Edith died at ninety, Gerda at ninety-six, and Lester at one hundred years of age.

Their long lives were so enriched by their experiences in California's wildlands and their engagement in their native-plant-related activities was so intense, fruitful, and productive that, studying their lives, a kind of melancholy sometimes steals over me. Oral histories include the last days, the time when physical limitations end exploration. Seed companies give up the ghost, nurseries are sold, the trails see us no longer, our places know us no more.

The other major focus of our winter section, the lessons of coarse woody debris, contradicts this message. "Nothing really ends; it is just transformed" is a well-known sentiment, but the study of coarse woody debris provides the details that give such a statement meaning. A down tree in the forest, even when no longer recognizable as a tree, still influences future events. While researching these three women, I saw the connections circle around many times, the serendipitous meetings as well as the near misses, the important friendships, and most of all, the unforeseeable longtime influences. Several times Rowntree, while crawling on hands and knees over a coastal dune or around a mountain boulder, literally came head-to-head with another creeping botanist, whose identity she immediately grasped, based on prior knowledge of who else shared this arcane interest. In turn, the other immediately identified her as well. While reading Lester Rowntree's works, many of us experience this sudden feeling of recognition, encountering aspects of a woman whose life mirrors parts of yours or parts of how you would like yours to be. Lester's "just do it" message inspired Gerda Isenberg's creation of her nursery, and Gerda inspired the present owner, Kathy Crane, to continue in her footsteps. Writing about Gerda Isenberg, I cannot avoid the personal reminiscence, since as Gerda's employee I knew her as a mentor and friend.

When I first learned that there was a woman who had been given the name of Bahai Wakidu, the Seed Seeker, I wanted to know everything I could about her. When I found how accessible and friendly the Mendocino County Historical Society was, how Lila Lee would support and encourage my delving into the life of Edith Van Allen Murphey, I was as excited as if I'd discovered a field of wildflowers, uninvaded by exotic species, open to my explorations. Edith Van Allen Murphey survived enough personal tragedy to provide material for several daytime TV series and yet always retained her interest in what was around the curve in the trail.

Indigenous peoples in many parts of the world devote significant energy to keeping vivid the stories of the past, certain that keeping the ancestors alive in this way is important for the survival of the people. Besides naming gardens (the Lester Rowntree Native Plant Garden, the Gerda Isenberg Demonstration Garden, the Theodore Payne Foundation), besides keeping books in print, bringing books back into print, and anthologizing articles, we can do one other thing—and for this I am grateful to the work of many, from Lila Lee at the above-mentioned institution to the skilled storyteller Rosemary Foster—we can continue to tell the stories. Let's begin with Lester.

12 Supreme Advocate for California's Native Plants

Lester Rowntree

I doubt that we'll see her like again on the Sierra trails.

James Roof, 1978

In 1977, newly employed by Gerda Isenberg as the seed propagator for Yerba Buena Nursery, I searched the shelves in a small back room for what I could lay my hands on of the native plant literature as it existed at the time. Here I discovered well-worn copies of Lester Rowntree's books *Hardy Californians* and *Flowering Shrubs of California.* These were autographed first editions; Gerda Isenberg knew Lester and was greatly inspired by her. In this she was not alone.

Well known in California native plant circles for her intrepid solo botanical explorations, her passionate dedication to the flora of California, and her seemingly effortless, vividly evocative prose, this strong-minded, unconventional woman was born in northern England in 1879 and lived to be one hundred years old. We know Rowntree in the main from her horticultural books, *Hardy Californians,* written in 1936, and *Flowering Shrubs of California,* written in 1939. We also know her through the work of several devoted biographers and a number of others writing about Rowntree in journals, who have kept her life and work continuously before us. Though both books were out of print for many years, and the hundreds of articles she wrote for a variety of venues over a period of thirty years remain uncollected, she continues to be an inspiration and a force in native plant horticulture and native plant conservation.

Today, it would be hard to find a professional prominent in the field of native plant horticulture who was not, at some point, inspired by Lester Rowntree. The model of her double focus, wildland exploration and landscape use of native plants, is followed by numerous California native plant horticulturists, from arboretum directors to landscapers to nursery professionals, who make regular trips into the wild for the pleasure of observing plants in their homes and to collect seeds and cuttings for propagation.

ROWNTREE'S BIOGRAPHERS

Not only was Lester Rowntree a deeply informed, skillful, and eloquent writer, but her life and her work also drew an impressive eloquence from the many who wrote and talked about their interactions with her: fellow native plant enthusiasts, botanists, grandchildren and daughters-in-law, companions in later life, friends, and acquaintances. Rowntree fed our hunger for those who, living long, retain the passions to which their years have been dedicated. Her midlife eruption into freedom to follow her dreams is frequently cited by those for whom her optimistic commitment rings the chimes of adventure and fulfillment.

Articles about Rowntree abound. Three women, Rosemary Levenson, Skee Hamann, and Rosemary Foster, have written or lectured about her extensively. Rosemary Levenson was responsible for the Bancroft Library oral history written in 1978 and the introduction to the reprint of *Hardy Californians* in 1980. Her exhaustive interviews for the oral history include those who knew Lester in a variety of capacities, from family members, caregivers, and personal and professional friends. She left no stone unturned to produce a complete portrait of a century-long life, and her oral history is equally a study of aging, with both its undeniably grim hardships and its possibilities for reconciliation and insight. Skee Hamann wrote "Runaway," a biography of Rowntree, as well as her biography in *Fremontia,* the journal of the California Native Plant Society.

More recently, many of us know Rowntree through the entertaining, informative lectures of her present-day historian, Rosemary Foster. Through years of searching in horticultural libraries throughout the United States, Foster has compiled an invaluable bibliography of over 720 journal and magazine articles by Rowntree, beginning in 1928 with a flora of the Carmel area and ending in 1977 with two detailed articles published in *Fremontia* only two years before Rowntree's death.

LESTER ROWNTREE AND COMPANY

Gardening and love of wildflowers was an early part of the life of Gertrude Ellen Lester, who always bore a deep and abiding affection for Penrith, Cumberland, her birthplace in northern England. To her older grandson, Rowan Rowntree, she said: "I drool with nostalgia when I think of England."

At the age of ten, Rowntree moved with her family to the plains of Kansas,

FIGURE 1. Lester Rowntree with plant press, 1936. Photo courtesy the California Academy of Sciences.

where they lived for two years until moving to a Quaker community in Altadena, California. Here she fell in love with the wildflower fields of southern California. At age twenty, she enrolled at a Quaker school outside of Philadelphia, where, in keeping with the custom of calling students by their last name, she came to prefer being called "Lester."

In 1908 she married Bernard Rowntree, also of English Quaker descent. Outside their fine home in Oradell, New Jersey, Lester created a garden so beautiful that, in the time-honored English tradition, she opened it to the public. In 1911 their son Cedric was born.

After contracting an illness that she was told would be terminal, the Rowntrees moved to California, first to southern California and then to Carmel. Lester's wish was to "expire amidst the splendid wildflowers of southern California," yet their curative powers may have been such that, instead of expiring, she began to thrive. With her son grown, and her marriage in a shaky state, Rowntree devoted herself to studying and beginning to write about the flora of California.

At the age of fifty-two, on the heels of a divorce, Lester Rowntree burst forth with a scheme that would be her modus operandi for twenty-five years. In 1929 she began Lester Rowntree and Company, based in her new home in the Carmel Highlands, with a partner, Lila Clevenger. While Clevenger stayed in Carmel cleaning and packaging seeds and mailing them to customers, Lester roamed the highways, dirt roads, and burro trails of the coast, mountains, and desert reaches of California, collecting seeds and plants. She drove off in a specially outfitted automobile stocked with plant presses, seed bags, shovels, a sleeping bag, typewriter, and burro food, which in a pinch she occasionally ate herself. The range of her wanderings, as well as the quality of her prose, shows up in the following passage from the preface to *Hardy Californians.*

> But before the season's climax is reached in the Sierra, I shall cross and recross the lower westward ranges, meet the same rivers again and again, skirt the sea, wander through the cupped beauty of the sand dunes, follow up dry water courses where cottonwoods and sycamores tap the unseen supplies of moisture, search the New-England-like meadows of the northwest, travel long trails through the foothills under the gray-green boughs of Blue Oak and Digger Pine, over a floor of bright green or soft tan, with the lavender-pink of high mountains showing beyond.

SEEDS AND WORDS

Her ambitious scheme, to be physically present for the life-cycle events of hundreds of plants throughout California, was thrice-motivated: to gather seed and sometimes cuttings for propagation, to make her better able to recommend proper horticultural practices through firsthand observation of the circumstances of each plant's home ground, and for her own deep pleasure. It was carried out without research grants, with no support from institutions or help from graduate students, but with the structure provided by her seed company, Lester Rowntree and Company.

The fulfillment of her seed list served as a letter of authority, giving her efforts authenticity, validity, and purpose. The freedom offered by answering to no one but her customers may have compensated for the lack of a sponsor or an academic support system. The interplay between horticultural purpose, the naturalist's unlimited curiosity, the business person's practicality, and the author's dedication to sharing experience, flowed through her. She was making a case, through her seeds, books, articles, talks, and landscape gardens, for the plants of California.

Rowntree searched them out wherever they might be, from the deserts to the mountains and everyplace in between. She scrutinized their ways and made deep acquaintance with them, returning again and again over a span of twenty-five years. Setting herself a rigorous scheme of study, she left her base camp, reached by burro or pack train, at sunrise, walked to the places where burros couldn't go, and spent the day and sometimes the evening with the plants. Sometimes it was a return to an old friend, sometimes a new discovery. She bore witness to relationships between insects and plants and to the changes through the day and the season, to the budding, the flowering, and the going to seed. All these were grist for her mill. Rural inhabitants sometimes found her unorthodox occupation peculiar, but to her it seemed "natural and plausible."

> To the conversant the procedure of tracking down flowers is natural and plausible. After the eager encounter must come a sojourn long enough to learn the day and night habits of the plants, to photograph, to press a specimen and perhaps take some seed, to study the symbiotic relationship between vegetation and its attending insects. Not until intimacy has been established can a move be made to the next stand.

The notes she made during her times in the wildlands of California formed the basis of her two published books, and she intended to publish at least three more books, partial manuscripts for which are held at the California Academy of Sciences: one on desert plants, another on rock gardens, and a third in diary form called "My Hillside Garden." A series of fires destroyed her seed collections and writing notes in 1949. According to James Roof, first director of the Regional Parks Botanic Garden, Lester was devastated by the fire. Roof says, "Losing all that work really broke her up. She was never the same again." Yet Lester went on to write a number of her best articles after those events and, with the end of her seed company, felt to some degree freed from the rigorous rounds it demanded.

One of Lester's stated goals in *Hardy Californians* is to prove to East Coast gardeners that the alpine plants of California might thrive in their gardens and that they are compelling enough to be worth a try. Aiming to convince gardeners outside California of their usefulness, she writes, "American horticulture would take longer strides if plantsmen in the East knew the Pacific Coast material better and if we on the Pacific Coast were more familiar with the plants and growing conditions in the East."

Rowntree is understandably sympathetic toward East Coast gardeners, since her first gardens were made in England and on the Atlantic seaboard. Gardening in New Jersey, she longed to try out California's flora and brought back her own seed collections for trials. She wrote many hundreds of articles for national gardening magazines, whose readers were mostly from the East Coast. But today, the main thrust of gardening with California native plants is to use them in California gardens. Whereas in Lester's early days, according to James Roof, there were only about ten native plant horticulturists in the state, the numbers of California native plant gardeners are now legion, and we no longer look to the East Coast as the arbiter of the worthiness of our native plants as horticultural subjects.

In *Hardy Californians,* we travel down the mountains with Rowntree, starting at the sublime peaks, where the plants of interest to rock gardeners and Eastern gardeners are described; we in the lowlands can only sigh in envy. Next she somewhat regretfully leaves the peaks and descends to the subalpine, where battered trees provide shelter. Below are the foothill regions.

By the middle of the book, Rowntree reaches the lowlands, and the rest of the book provides treasures for gardeners in the California Floristic Province, that slice of California between the ocean and the Sierra. She takes her material genera by genera, making verbal pictures with skill and her trademark playfulness. Her hardheaded and useful chapter on growing California annual wildflowers may be one of the first treatises on this subject, and it is still wise.

Rowntree's years of dedicated exploration of California's native bounty gave us one of the classics of native plant literature. The information she provides is still current, still invaluable, and should become even more so as the number of gardeners of the Sierran foothills increases exponentially in the next fifty years. Her goal of encouraging Eastern gardeners to use California's alpine plants is now overshadowed by the need for this information to reach a group that so far has not received its due from native plant literature, the gardeners of the Sierran counties.

Given current projections for population growth into the foothills, and given what we know about the results of uninformed horticultural choices and activities in the current major population centers in California, it behooves us to be ready for the next surge, which is gathering speed. We need to be proactive in influencing the decisions of the gardening public and help these communities avoid the severe consequences of mistaken plant introductions.

Montane and foothill native plant nurseries are in place, as are inspiring local demonstration gardens and landscapers equipped with the knowledge and focus to help home owners keep their gardens in the Californian framework. The native plant garden tours that have proved so effective in large population centers should be begun early. *Hardy Californians* would be an essential part of the plan. Rowntree's rich and inspiring text can be an effective tool in the protection and restoration of Sierra foothill plant communities threatened by bulldozers and subdivisions. Her descriptions of some now greatly changed areas may make us yearn for the lost beauty of an earlier time, but they also provide invaluable information that can give us important cues for foothill and montane restoration gardening and activities.

FLOWERING SHRUBS OF CALIFORNIA

Between *Hardy Californians* and *Flowering Shrubs of California,* which was published three years later, a significant shift toward promoting California plants for Californians is evident. The last chapter in *Flowering Shrubs of California,* "Culture and General Councils," is particularly focused on Californians gardening with California native plants. In 1936 she writes,

> Particularly during the past five years there has been in California an increasing awareness of our native flora and a stirring of interest among gardeners. People have begun to feel like having a try, anyway, at growing natives, especially when they hear that most natives get along without water, fertilizer, or insecticides. And this interest is awakening a response among nursery men; if the demand grows strong enough the supply must surely follow.

This stirring and awakening has happened several times in my lifetime as well, with native plant advocates citing the same factors behind the non-use of California natives in gardens: lack of cultural information and lack of availability. Each time, native plant aficionados become wildly hopeful. And each time the movement is strengthened, more plants are grown with more knowledge, more native plant nurseries produce more material, more good writing inspires and informs. Meanwhile other kinds of planting continue as well, and also in increasing numbers. In a place of staggering population growth such as California, the percentages may remain constant or may even shift somewhat significantly in the direction of native plant gardens, but the overwhelming picture will still be gardens that for one reason or another do not reflect, promote, or enhance our increasingly diminished native heritage. We need all the tools at our disposal to maintain and increase the gains we have made. Both of Lester's books are mighty instruments; they instruct while they delight, the more frequently read the more deeply enjoyed.

In the last chapter of *Flowering Shrubs of California,* Lester presents a cogent scheme for the planning and planting of what she calls "wild gardens." It reveals her knowledge of gardeners and the mistakes that can be made using native plants. She effortlessly describes a variety of approaches, beginning with the use of your local "weeds," those ubiquitous native species that you know will grow well in your situation, and recommends the beautiful back-and-forth play between wild garden experience and knowledge of wild species in their homes that is at the heart of Rowntree's approach to native plants.

Numerous specific references to the occurrences of different species and plant associations throughout the state of California should prove valuable to those seeking to understand changes in the landscape. In her chapter on chaparral in *Flowering Shrubs of California,* besides an informative and amusing background on chaparral, she tells us where to find it.

> There is good chaparral up Bouquet Canyon in Los Angeles County, and you will find excellent stands in the backcountry of San Diego and Ventura counties, up the Cuyama River in Santa Barbara County (especially on the north facing slopes), and in San Luis Obispo County. Or take the Angeles Crest Road to Mount Wilson.

THE THINGS THAT SHRUBS HAVE TOLD ME

Unlike John Muir, with whom she is often compared, and who didn't give much thought to gardeners, Lester traveled with a gardener's focus, considering which plants could be "displayed with pride" in the garden, and which were "insignificant" or of little interest if not particularly showy or "garden-worthy." Lester Rowntree as a pioneer singing the praises of the California species encountered the necessity that many professionals encounter, that is, the need for native plants to match in showiness, size of bloom, and intensity of hue the flora of South Africa, Australia, Greece, and Chile, other Mediterranean climates. Native plant gardeners of today are frequently ecological sophisticates, aware that every species has its place and its value. But the need to compete with the worldwide flora is even greater than before, and the way into many an uninitiated gardening heart is through bedazzlement. This Lester well knew.

Lester knew plants in two ways, from observation in the wild and from growing them herself.

> I have put down what I have gleaned from personal observation of the habits of these shrubs in the wild and of the behavior under cultivation of those which I have grown in California during the last 12 years. In compiling it I have followed my invariable rule of writing only from my own notes, taken on the spot, of the things that shrubs have told me in personal interviews.

Lester's description of Pacific wax myrtle, *Myrica californica,* is an example of the concision of her prose. Because of its usefulness as a fast-growing, long-lived hedge plant, it is currently enjoying a comeback, being returned by gardeners to many places where it used to grow. Lester follows it from its beginnings in the Santa Monica Mountains just north of Los Angeles all the way up the coast to Oregon. Along the journey, she describes its variation in height and width and growth habit, the way it takes on and leaves behind companions, the different situations in which it grows and how they affect its appearance. She sums up its cultural requirements in one succinct sentence. "Grow *Myrica californica* in light high shade, unless you live in a foggy belt, and in rich humusy soil, and don't be afraid of overwatering it in summer." Here it is, all you need to know about growing *Myrica californica.*

LESTER AND THE AUTOMOBILE

Lester's relationship to automobiles foretells the tricky, potentially disastrous nature of this beast that those of us further along in the saga know all too well. Continually extolling the virtues of plant exploration where cars cannot go, she yet depended on them, as who does not, to get to the trailheads and packers and burro owners. Frequently defeated by unreliable vehicles, locking herself out of cars, experiencing flat tires, vapor lock, and other dispiriting ailments in remote reaches, and once even barely escaping from her car as it teetered and then finally descended over a cliff, she tolerated cars as the tedious but critical means to her glorious ends.

She felt that she operated under a disadvantage: "Modern boys and girls are born with an instinctive knowledge of cars which I lack entirely. All that I know has been learned by hard experience. . . . However, I have come to think there is a guardian angel for car-ignorant wandering women. The angel generally employs Highway Division or Forest Service men as instruments."

James Roof describes her frustration with automotive breakdown in the Central Valley in straitened economic circumstances and the freedom and joy she experienced when her first book advance was turned into a reliable vehicle. Rowntree paints vivid pictures of her restructured automotive interiors, customized for the purposes of her occupation. Swiveling driver's seats allowed for a range of activities other than merely driving, and the removal of passenger seats freed the space for other, more critical functions. Grandsons and other passengers, when necessary, sat on overturned buckets.

The tools of her trade are lyrically limned in many of her articles as well as her books, a hymn to camera parts, plant presses, newspapers, and juice cans. Ingenuity and innovation are her themes.

ROWNTREE'S WRITINGS FOR CHILDREN

That Lester had a particular interest in the lives of rural children is evident in her four books written for children: *Ronnie, Ronnie and Don, Little Turkey,* and *Denny and the Indian Magic.* In these books, young boys have magical adventures with Indians or meet up with old-timers who know important and necessary things about the land. Though now out of print, these books are unrecognized California children's classics, of which there are all too few, books that can help our children feel what it means to grow up in close association with the California landscape. They are cultural histories of California's country settlements, contributing to our sense of California's early rural life, which is poorly documented compared to that of New England or the rural South and Midwest. She captures pungent vignettes of backcountry and high country life, spending as much time as she did, on "earthy country roads upon which amiable local people go slowly about their business."

Little Turkey is particularly rich in rural lore. We learn how an early homesteading family in the foothills of the Sierra ate and lived, farmed and ranched. Native plant lore is interwoven throughout, as the hero, the shy, tough character Little Turkey, travels through fields of lupines, revels in monkeyflowers in creeks, uses pine tree resin for drawing out splinters. A "whiteheaded woodpecker" lives in a tree by the kitchen. The busy, unceasing ministrations required to raise little ones by both the human and the avian mother take place side by side. As Little Turkey walks through the woods on his way to school, he observes a blackbird perched on the back of a deer. Apparently this sight was so common as to merit no particular comment, just one of the sights that might frequently be observed on the way to school. It is assumed that the bird is removing insects and ticks. In 1944, J. S. Dixon reported on a similar phenomenon: a western scrub-jay grooming a buck. In both recountings, the deer were described as offering themselves with pleasure and gratitude to the ministrations of the bird. African mutualistic relations between the rhinoceros and the oxpecker are better known than our Californian versions. Rowntree captured many such gems.

RAMBLES IN THE WILD

Horticultural endeavors, and the gardens in which they take place, assume innumerable forms. They range from the harmonious, graceful, novel, useful, edible, and medicinal to the endangered, showy, and bizarre, embodying the full panoply of human values. Lester's horticultural impulse rose from ecstatic solitary journeys, daring rambles that not infrequently courted danger. In one incident, as noted above, her car went over a cliff. Though she wrote about this and other incidents with her unique light touch, as though it were a humorous incident, her only concern the bouquet she had gathered for a talk she was planning to give, the reality remains that she took risks, and not insignificant ones. Encounters with rattlesnakes are a recurring theme in many of her short pieces, as well as in both books.

Rowntree was a pilgrim looking for beauty. She wanted to make it accessible, save it, honor it. The relationship between gardening and wildland preservation is one that Lester Rowntree helped illuminate. Without addressing these issues specifically in *Hardy Californians,* though elsewhere she did, she makes the case for protecting wildlands and undertaking efforts to control invasive weeds. The motivation is implicit in the descriptions of the many stunning and vulnerable wild gardens she finds and encourages you, the hopeful native plant gardener, to re-create.

By describing with such love these plants and the places where they grow, she promotes their protection and survival. In this she resembles John Muir, whom she captures with a few brush strokes, in the context of describing alpine *Senecio:*

> One of the best *[Senecio]* is *S. muirii,* named for that famous pedestrian who, armed only with a tin cup, some matches, bread, tea, and (on a feast day) raisins, stalked through the Sierras in rapture and left us such perfect accounts of his visions.

Like "that famous pedestrian," Lester's "accounts" of her travels leave us more knowledgeable, more wistful, more inspired, and more determined. With sharp tongue and perceptive judgment, she assesses the plants she hopes to beguile us into using in our gardens. A fair number of the species Rowntree describes in *Hardy Californians* that were obscure at the time are still not available in nurseries. Some are Sierran species not able to survive in the lowlands because they cannot handle nursery conditions, including the confinement in containers that makes summer irrigation necessary for plants that in the ground would remain dry. Others have been deemed unsatisfactory garden subjects, while still others we simply haven't gotten to yet. Some that she describes are like a call to arms to the native plant horticulturist, filling us with some degree of eagerness to bring them "into the trade."

THE MYSTIC WALKED WITH A PRACTICAL OVERLAY

In *Hardy Californians,* Rowntree went to the mountains with a practical purpose. She gave herself a reason to be out there; she was doing it partly for us, her customers and readers, including the gardeners of temperate regions where plants must survive extreme conditions, the rock gardeners, and anyone curious about the occurrences of different species and plant associations throughout California. The seeker searched, to add to a list in a catalog, to observe profoundly in order to give concrete advice about soils, exposure, and moisture, and as she put it, to "collect herself." And she left us a litany of beauty, gorgeous plain songs, Gregorian chants of carefully articulated prose. There's something here for us, a kernel, a nugget of truth about being human. The hunter or the seed gatherer knows that to seek animals or plants for practical reasons does not diminish reverence for these beings, or awe in their presence.

In the last two decades of her life, when she had to hang up her traveling

FIGURE 2. Lester Rowntree at age ninety-four.
Photo courtesy the California Academy of Sciences.

shoes, Rowntree turned her focus back to her garden, putting into it the energy and devotion her field trips formerly absorbed. She describes gardening from dawn till dusk, and the garden took her back. She found the truth in the assertion of Sydney Mitchell, a well-known horticultural writer of the time and fan of Rowntree: "One of the greatest advantages of gardening is the fact that it is a lifetime pursuit in which you are always a participant. You are never forced into the grandstand as physical vigor declines; your interest and knowledge increase with the years."

As native plant horticulture and wildland exploration informed each other in her life, one giving value to the other, each stage of life brought a different emphasis, a different focus. In the wild, she thrilled to see "wild floods of blossoms I used to cultivate in an English garden, those baby blue eyes and poppies which in California spill across the grass." In the garden, she treasured the plants that reminded her of Roof's "field memories."

Rowntree died in 1979, surrounded by her family, including her two devoted grandsons. Among the many young people she influenced must be included Rowan Rowntree and Lester Rowntree, who both chose professions in the fields of natural history. In 2006, they undertook a labor of love, the reprinting and updating of their grandmother's first book, too long unavailable. As well as enlisting Allison Green to update botanical names in *Hardy Californians,* and including Rosemary Foster's bibliography of Rowntree's writings, they added their own personal biography and rich remembrances of Rowntree.

I recognized from delving into Rowntree's life and influence that, for students of California's native plants, there is never enough time. Rowntree wanted a twin, one who could be in the mountains when she was in the desert, or working in the garden when she was in the wild. A sobering moment arrives, when we realize what it actually means that we are given only one life. For students of California native plants, it may mean field trips that we will never make, valleys that will go unexplored or to which we will never return, high peaks that will remain unknown. In her one hundred years of life, Lester Rowntree fully inhabited many different Californian ecosystems, from the coast to the desert to the redwoods and the chaparral to the Sierra and the foothills and back again. As much as anyone could, she lived California.

13 Bright Were Her Days

Edith Van Allen Murphey

Edith's compulsion, which stayed with her till she died, toward gathering seed from an ample stand to scatter elsewhere, led the Shoshone to call her Bahai Wakidu, the Seed Seeker.

Skee Hamann, "Edith Van Allen Murphey," 1980–1983

While Lester Rowntree roamed the mountains, desert, and seacoast, another native plant lover was collecting seeds and plants in the mountains of Mendocino County. Born the same year as Lester Rowntree, Edith Van Allen Murphey was another woman ahead of her time in her love of California's wildlands, her courage in undertaking regular solitary expeditions to collect seeds and plants, her devotion to the plants of California, and her unusual (for the time) marital history.

I first encountered Edith's book, *Indian Uses of Native Plants,* in Gerda Isenberg's library, at Yerba Buena Nursery in Woodside, California, in my early seed-collecting days. Shortly thereafter, I found Skee Hamann's beautiful and concise article on Edith in *Horticulture* magazine, "Seed Seeker of the Flowering West." I was heartened by the example of another earlier seed seeker. Later, when my seed catalog expanded to include books, *Indian Uses of Native Plants* was one of the first books I carried. Twenty-five years later I carry it still.

Lester, a prolific and polished writer, believed in rewriting, while Edith, less organized and folksier, refused to, as she put it, "bake the same dough twice." Yet both were able to convey through their writings and through their lives the extraordinary nature of their experiences. They were known to each other. Among Edith's papers, I found a well-thumbed copy of an article by Lester Rowntree, "A Day among Californian Wildflowers." Rangers in the Yolla Bolly wilderness frequently told Edith Murphey about Lester Rowntree, but they never met. Though both were devotees of Willis Jepson, collected specimens for Alice Eastwood, and even shared a biographer in the form of Skee Hamann, they moved in very different circles. Unlike Rowntree, who claimed to have slept at least one night in every state in the union, Murphey, except for ten years when she ranged widely, working as a range botanist on Indian reservations in the intermountain West, spent most of her adult life in the relative wilds of Mendocino County, removed from botanic gardens, horticultural societies, and even the legendary "ten people in the state interested in native plants" with whom Rowntree fraternized in the 1920s and 1930s. Instead, she had direct and continuous contact with indigenous peoples, absorbing from them information about California's flora and the ways it sustained life. Where Rowntree wrote for numerous national and state journals, and her books found national publishers, Murphey wrote columns for local newspapers, and her one book, which appeared when she was eighty, was published by a historical society. Yet their long lives had in common a richness of accomplishments, with its basis in California's flora, that made them sisters. They were desert lovers who delighted in sleeping out in the desert alone. They loved the high peaks and used stock animals to help them get there. They collected with a strong awareness of the need for conservation. Others were frequently startled to see them pursuing their solitary ways. Both had a home place in California that drew them back from travels elsewhere.

In the years when Hamann was shuttling back and forth between these two aged seed seekers, both Rowntree and Murphey became more and more interested in meeting each other. From Hamann's correspondence, we learn that a visit had been arranged, but just before it took place, Rowntree became ill and the visit was postponed. By the time she recovered, Murphey had died.

As Hamann said of Edith, "For two years (before her death) I had visited her for interviews, and when she died she willed to me her botanical files and her unpublished writing. Each time I return to consideration of the vitality and purpose that propelled her through 90 busy years, I am as impressed as I was at my first knowledge of her life's fullness." Murphey was featured on two national radio shows, Mary Margaret McBride's interview program in 1954, and Edward R. Murrow's *This I Believe* in 1955. For neither of these interviews could she pry herself away from local events to go to New York.

"I KNEW THERE MUST BE ACCOMPLISHMENT"

Like Lester Rowntree, Edith Van Allen Murphey (1879–1968) came from a genteel background, and like Rowntree, an adventurous spirit claimed her early on. Though born to a well-off old Dutch family in Albany, New York, Edith's childhood was haunted by tragic deaths, including first the death of her mother during Edith's birth, then the deaths of two brothers from diphtheria, and finally her father's death when Edith was fourteen years old. Yet she retained a sunny, optimistic, accomplishment-oriented perspective on life. ("I especially was told that my life must be such as to justify the sacrifice of my mother's life. I knew there must be accomplishment.")

Her tomboy nature encouraged by her father, Edith reveled in the ancestral lands of her Slingerland family; yet she came to realize that the 10,000 acres bought from the Indians by Tunis Cornelieuse Slingerland in 1653 might not have proved such a good bargain for the Indians. Her whole life was fueled by a sense of injustice to Native Americans, culminating in her attendance, several weeks before her death, at a council meeting challenging the State of California's plan to flood Round Valley in order to provide water storage for southern cities. Early exposure to the Indians of Montreal and many stories from her grandmother piqued her strong curiosity and affinity toward native peoples. Much influenced by her father's love of books, and denied a career in medicine by the qualms of her protective relatives, at age twenty-one Edith attended the New York State Library School. Here she had frequent conversations in the elevator with the then governor of the state, Theodore Roosevelt, who encouraged her interest in Indians and the West. At the age of twenty-three, the lure of a job as a librarian at the University of California in Berkeley, with the promised proximity of wild land and Indians, drew her to make a solitary journey across the country.

Living in a boarding house in Berkeley, Edith connected with a group of young people who dreamed of homesteading in the wilds of Mendocino County, much as young people were to do sixty-five years later. In 1903, in a daring move for a gently reared girl of those times, she and two women friends took out homesteading claims near Willits. Here, according to Edith, they might have starved to death were it not for the aid of local Indians. What she learned from them was incalculable, and the friendships she formed with them lasted her lifetime. Edith belonged to the relatively small yet significant group of white settlers who moved in the Indian world with appreciation and respect, both witnessing and trying to ameliorate the devastation of Indian lives. In this she was far in advance of some of her peers, including John Muir, Carl Purdy, and Louise Amelia Knapp Smith Clappe, author of *The Shirley Letters,* but in the company of a select few, such as Jeff Mayfield, author of *Tailholt Tales,* Joaquin Miller, author of *Life amongst the Modoc,* Mary Ellicott Arnold and Mabel Reed, two schoolteachers who wrote *In the Land of the Grasshopper Song*

FIGURE 3. Edith Van Allen, circa 1900. From *Bahai Wakidu: The Seed Seeker.* Courtesy of the Mendocino County Historical Society.

after teaching on the Karuk reservation, and notably Effie Hulbert, author of *Indian Summer,* published by the Anderson Valley Historical Society. This last is an unrecognized treasure, written by a woman who grew up with a band of Pomo living on her family's ranch in Anderson Valley in Mendocino County, partially protected by them from the incursions of the white world. When her family lost the ranch, the Indian families she knew intimately became essentially homeless. Effie, like Edith, spent her life trying in some way to make up for what she perceived as injustices and as the destruction of a valuable way of life. Her manuscript includes one shining page on her memories of her native friends preparing and eating native plants.

In a well-preserved Victorian house in Ukiah, I first delved into Edith's letters, articles written for a local newspaper in Round Valley, and, most important, a manuscript that was meant to be a biography of Edith, written by a woman who knew her well and was her devoted friend, Skee Hamann. I also encountered Skee in Lester Rowntree's archives as the author of "Runaway," a short biography of Rowntree. Unfortunately, Skee died before she was able to see her book on Murphey through to published form, but the facts of Edith's remarkable, dramatic, accomplishment-packed life are handsomely delineated in the two versions I examined. As I turned the pages of the first manuscript, I exclaimed to Lila Lee, an indispensable longtime volunteer at the Mendocino County Historical Society, "She got married again!" Then a few chapters later, "She got married again!" It was a story as riveting as any soap opera.

Edith first married the homesteader who owned the property next to hers, a Harvard graduate named Arthur Lloyd. They built a simple cabin in the redwoods, graced by a gilt clock from France, sent by her family in Albany. Deer and rabbits destroyed their first two gardens, and starvation loomed until their Indian friends took them in hand and taught them the skills they needed for survival: how to spear and smoke salmon, how to make deer jerky, and how to eat the local native plants.

> Many of the Indian native plant foods still survived in Mendocino, Blue camas bulbs, the bulbs of the numerous species of brodiaeas, the leaves of the yellow cowslip, the Miner's lettuce and water cress for green stuff, the wild onions; tangy bulbs, the wild berries, the sweet rich pinenuts, grass seed for pinole. (35)

Edith reveled in this existence. She described to Skee their first idyllic Christmas in the little cabin, snow coming down, fireplace aglow, a little tree decorated with strings of rosy-red toyon berries, her husband's slippers warming by the fire. Outside a winter storm raged, as Edith waited for Arthur's return with supplies and Christmas mail. He arrived by horseback, bringing with him a package from the Albany relatives, containing a layette in exquisite wool, linen, and lace for their anticipated baby. Also provided was "a Tiffany rattle of mother of pearl set all about with silver bells." The layette was carefully laid in an equally valued willow baby basket made for them of silverleaf willow *(Salix hindsiana)* by a Pomo friend, Steve Sherwood. This poignant image, the elegant baby clothes and blankets lying in the finely made Pomo basket, vividly represents the gift that Edith brought to her new home, a lifelong openness to and respect for indigenous ways. In this small cabin in the woods, Edith's layette for her awaited child is a symbolic and actual meeting of the ways two cultures greeted newborns. The consequent tragedies are all the more poignant for the perfection of this moment. As Edith said, "For a time, fate let us be happy."

When their daughter was fourteen months old, they responded to a neighbor's call for help with a difficult birth. During a violent winter storm, with their daughter in the saddle in front of her father and Edith on a young colt, they set out to help. While fording an overflowing creek, Edith's colt lost his footing, was swept away, and drowned. Edith, who was almost nine months pregnant, barely got her feet out of the stirrups in time; a Pomo Indian who was gigging salmon nearby speared her coat sleeve to the willows and saved her. They made it to their neighbor's house, helped his wife give birth to a healthy baby boy, and returned home. Edith herself shortly went into premature labor, perhaps initiated by this series of events. Her baby did not live.

The idyll began to break apart. Their daughter, Margaret, developed a "violent emotional illness," possibly severe autism, possibly epilepsy, and was taken to the state hospital. Edith's Albany family sent her older sister to help the bereft couple. In an unfortunate incident that was a bitter memory throughout her life, Edith discovered her sister and her husband in bed together. She left the homestead immediately and never saw it, her husband, or her sister again. In her later years, twice widowed, Edith was contacted again by Arthur Lloyd, then living in Los Altos, who described his lifelong regrets and proposed that they meet. Unable to forgive, Edith refused. Yet all her life she kept on her wall a picture of their cabin painted by her husband, "warm redwood timbers, touched by sun, sheltered closely by leafy green oak and madrone."

In 1906 Edith married Sanford Redwine, from a Round Valley ranching family, and for eleven years they were happy together, during which Edith continued her self-taught botany studies, her investigation of indigenous ways, and her basket collecting for the Phoebe Hearst Museum. They made their home in Covelo, where Edith returned time and again during her ten years of extensive travel, and where she died. At age fourteen, Edith's daughter died in the mental hospital where Edith had gone to work in order to be near her. Though her death released Margaret from twelve years of mental anguish, Edith was devastated. "The poignancy of the short happy months of Margaret's normal infancy, stabbed Edith's heart in unguarded moments all her life, reminding her of 'A homestead cabin, a fireplace, a sleeping girl baby, and the peacefulness of life in the hills.'" Shortly afterward, Redwine, who suffered from tuberculosis, shot himself, the note he left behind reading that he could not bear to be a burden on Edith any longer (50).

Hamann describes Edith's way of handling the tragedies that life kept hand-

FIGURE 4. Edith Van Allen and Arthur Lloyd's homestead cabin near Longvale, circa 1905. From *Bahai Wakidu,* Mendocino Historical Society. Oil on canvas; painted by Arthur Lloyd. Black and white reproduction courtesy of Donald Winters.

ing her: "The hurt of her grief was as deep as ever but she had learned to live above it in the intensity of her concern for others. The 'stab in the heart' was not allowed to stay on the surface of her consciousness. Vigilant mental discipline forced it down by concentration on constructive busyness" (87).

Alone again after her third husband, rancher Will Murphey, died of sunstroke, Edith threw herself into plant study, mastering Jepson's *Manual of the Flowering Plants of California.* It was then that Ace Hoagland, a young Mongothl Indian, came to help her at the lonely ranch, and she became intrigued by his "seemingly inborn knowledge of Indian uses of native plants" (52).

After selling the ranch where she had lived with Murphey, she found herself, at a vigorous age forty-eight, alone and without responsibilities. Like Lester Rowntree, she looked longingly at the high country, possessed with the desire to spend time with the alpine plants. Also like Rowntree, she became a collector of seeds, bulbs, and plants. Carl Purdy of Ukiah, who supplied California native seeds and bulbs to European gardeners and nurserymen, hired Edith to cook for him in the winter and collect for him in the summer. On these adventures, she included Ace.

Edith waited with anticipation for the signs that the high country was accessible.

> When spring comes and feed begins to grow in the mountain meadows and my friends tell me that the red bells are blooming just below the patch of snow in the low gap, Glory and Mandy will come in off the range to be fitted to new shoes, my camp outfit will be overhauled and as soon as we are reasonably sure of settled weather, Ace and I will start out again. (57)

They would head for the as yet botanically unexplored Mendocino National Forest above Round Valley, land that, coincidentally, the friend of her youth Theodore Roosevelt had designated as public land. All her life, the lure of the wild country energized her.

> The one sound that stirs my blood above all others is the far, faint call of wild geese high above me in their V-shaped wedge heading north because spring has come. The lure of new country—the call of adventure—is in my ear. Stronger than usual I have my chronic feeling that just over the hill, around the turn, something new and interesting is waiting for me. (56)

Like Rowntree, she had her adventures with rattlesnakes, but unlike Rowntree, at least for this time, she had "the ideal camp companion" in Ace, who took care of the horse and mule, cooked, smelled out rattlesnakes, and oversaw the gear, while Edith checked fieldnotes, seeds, bulbs, and flowers in the presses before she began a new day. As well as collecting for Carl Purdy, Edith supplied dried plant specimens for Willis Linn Jepson, Herbert Mason of the University of California Herbarium, and Alice Eastwood at the California Academy of Science in San Francisco. She and Ace roamed the mountains of Mendocino, from wet meadows to gravelly ledges, to snowmelt slopes and rocky plateaus.

"BRIGHT WERE OUR EVENINGS OF SONG AND STORY"

From Ace, Edith learned to replant the outer scales of the bulbs she collected, to avoid harvesting in the same spot in successive years, and to abstain from taking both seed and bulbs in the same location. Edith began to incorporate Indian ways of sustainable harvesting into her seed-gathering activities. Ace taught small Edith, whom he called Big Boy, how to cook at high altitudes in earth ovens and how to use the forest resources for medicine, bait, and fiber. They had a sociable time of it.

> Ace built a big supper fire, for it is cold in the high hills when the sun goes down. Bright were our evenings of song and story. Ace insisted on hearing me sing. His favorite was "Hot Time in the Old Town Tonight," and when I suggested that no one had ever wanted to hear me sing before, he said earnestly, "That's all right, Big Boy. I know all your songs got the same tune. I just want to learn the words, that's all." (60)

By the time Ace grew out of childhood, Edith had the skills she needed to explore the mountains by herself. Like Rowntree, she startled rural inhabitants unaccustomed to seeing small, middle-aged women roaming the wilderness alone. To some, this divorced, three times married woman in trousers was outside the pale of respectability. Others watched out for her and tried to help. Edith also collected seeds for Lester Rowntree and Company. "We had much correspondence. I read her books and envied her the photographs of flowers. I always wanted to hear her lecture but we never met" (64).

THE SALT JOURNEY

In 1927 Edith became close friends with Lucy Young, a Trinity Wintu Indian (Lasski-Wailaki tribe) born around 1843, who lived to be 102. Lucy Young was a major influence in Edith Van Allen Murphey's life. She thought of her as a "wise woman" and treasured what Hamann called their "disciple-sage relationship." Young's memory of a time before the arrival of white people, the many tragedies she had survived, reminiscent of Edith's own sorrows, her remarkable oral history of her people, and her plant knowledge—from all these Edith drank deeply. "Indians came to her for plant and game locations. She had only to sit awhile in the sun against her cabin wall, with her eyes closed, to be able to tell them where to search" (76).

Edith recorded Lucy's stories, combining them in an invaluable document, published in several venues. These stories include truly horrific accounts of the treatment of Indian children during the gold rush and the dire prophecies of her grandfather, Yalleye. "My grandpa never live to see white people, just dreaming every night 'bout them." Not pleasant dreams, either.

These painful stories are yet replete with valuable details. One comment illuminates the impact of overgrazing by cattle, followed by the invasion of weeds, in fields where Indians formerly gathered nourishing bunchgrass seed. Lucy was a firsthand observer, she experienced and understood the events her grandfather predicted, the biological turnover from bunchgrasses to weedy alien grass and weed species considered by many to be one of the quickest, most devastating in history. Lucy Young felt its impact in her own hungry stomach.

When Lucy was ninety, and her husband eighty, they decided to take Edith on a two-week pack trip to South Fork Mountain. "We had talked for months of a pilgrimage to South Fork Mountain to find medicine and basketry plants, see the Shasta lilies 'in sheets' and let Lucy live for me the long ago journeyings of her people" (78). This journey would remain one of the most memorable experiences of Edith's life. As they rode, Lucy recounted stories of the salt journey her people made each year to the salt springs on North Yolla Bolly. "It was a two-month journey on foot, idyllic, as they traveled leisurely through country where every woods, glade and stream was associated with tribal history and legend. On the way in they marked food bulbs in bloom, such as Indian potatoes, Indian onions, Camas, for digging when ripe on their return." On this journey Edith practiced the lessons that Ace had taught, collecting only what they needed to eat, leaving some for others who might come by. "Seeds and bulbs were never gathered in the same patch" (82).

If Rowntree's favorite plant was *Eriogonum ovalifolium,* Murphey's may have been the Shasta lily (known then as *Lilium washingtonianum* var. *minus,* now *Lilium washingtonianum* subsp. *washingtonianum*). This extremely fragrant, waxy white, high-altitude lily has lost much territory to logging. Though usually not successfully cultivated, it has been extensively commercially collected. "For searchers, the impact of sighting its magnificence is a bit like finding the Holy Grail" (84). Not a little of the impact of the experience would be scenting the flower before actually sighting it. On this trip, they were successful on Horse Ridge in Trinity County:

> On the afternoon Lucy and I rode up to Horse Ridge, the ground was covered with Pussy Paws, watermelon pink; on a higher Ridge the bright rose shrubby penstemon, Mountain Pride. Above them rose what I had waited so eagerly to see. Glistening white Shasta lilies, in sheets, sticking their heads through the low oak brush on all sides, saturating the still air with sweetness. (84)

Though Edith is most well known for *Indian Uses of Native Plants,* she also wrote a small pamphlet called "The Stockmen's Pocketbook," based on her days as a range botanist for the Indian Service. For ten years Edith was employed by the Indian Service, working on eleven different reservations throughout the West. Everywhere she went, she encouraged native crafts and involved the children in making local floras, putting together wildflower shows, and collecting herbarium specimens. On the Blackfoot Reservation, she helped the children put together a pamphlet titled "Materia Medica of the Blackfeet." She was also employed by the United States Plant Bureau to record Indian uses of native plants in Nevada.

The uncollected newspaper articles she wrote during her professional days and also after her retirement figure heavily in Skee Hamann's biography. They capture much of the flavor of Edith's personality, as she summarized the lessons of the years, her thoughts on aging, and her love of the inhabitants and natural occurrences of Round Valley, where she ended her days, about thirty miles east of her first homestead.

When I had the opportunity to buy a small tract of land in Mendocino's third-growth forest to use as a "field station" for my entry into forest restoration, I thought frequently of Edith's experiences. But I learned only through the excellent footnotes of *Bahai Wakidu,* the Mendocino County Historical Society biography of Edith Van Allen Murphey, that I now share a watershed with Edith's original homestead. I could possibly walk to the creek in which Edith almost lost her life, and if I do, it will be to think about how Edith Van Allen Murphey turned that life into one long giveback to the world.

Edith's years with the Indian Service, "botanizing joyfully around the West," were important, service-dedicated ones. She worked with a dizzying number of organizations on a stunning number of projects, meeting the perceived needs of indigenous peoples and enhancing their pride in their knowledge of native plants, while recording all she could. Spurred on by Lucy Young's terrible stories, she worked tirelessly with Blackfeet, Shoshone, Warm Springs, Paiute, and others. Yet the images that remain with me most powerfully are of her California life, including the photograph of Lucy and Sam Young on horseback with Edith during their trip to South Fork Mountain, a trip such as no other white people were privileged to experience. I envisage the fine baby clothes from France, laid with anticipation in the equally exquisite Pomo baby basket made of silverleaf willow.

14 An Inordinate Number of Good Things

Gerda Isenberg

I go to books and to nature as a bee goes to flowers—For a nectar that I can make into my own honey.

From Gerda Isenberg's bookplate, designed by her daughter, Ami Jaqua

In 1955, at the age of fifty-five, Gerda Isenberg founded Yerba Buena Nursery, which still exists today as one of California's largest and oldest native plant nurseries. She actively ran the nursery until 1995, one year before her death at age ninety-six. During those forty years, sixty-two employees and interns benefited from the unique situation that Yerba Buena Nursery provided. Located at the end of a challenging dirt road, on forty beautiful acres in the then relatively wild Santa Cruz Mountains, the nursery supplied ocean views, vernal ponds, venerable oaks, chanterelle mushrooms, a well-trained nursery dog and the opportunity to know and work with Gerda Isenberg. Many people, myself included, felt that working at Yerba Buena Nursery for Gerda Isenberg was a great stroke of luck.

As Bart O'Brien, the director of horticulture at Rancho Santa Ana Botanic Garden, put it, "Yerba Buena Nursery has long been an important source of native plants for the Bay Area, and an invaluable way station in many lives." Its large and well-grown inventory of native plants continues to be an unmatched resource for home owners and professionals interested in the flora of California.

O'Brien is among those who were employed by Yerba Buena Nursery in the forty years of Gerda's tenure; others include two former presidents of the California Native Plant Society, a wholesale fern grower, several ecological restorationists, an herbalist, many landscape designers, the landscape architect who designed the innovative San Mateo County development called Portola Valley Ranch, groundskeepers for estates and universities, a horticultural therapist, a permaculturist, estate managers, naturalists, a research botanist, a seed person, several writers of books about native plants, and a Buddhist novelist.

Gerda's influence is felt not only by those who knew and worked with her but indirectly by the many persons in this field who were educated or inspired by Yerba Buena Nursery's alumni. I speculate that many people currently involved with native plants, in whatever capacity, if they did not come directly under Gerda's influence, were influenced or educated or inspired by somebody who did. Gerda is the supreme example of the rock cast into the pond whose ripples go on and on.

Because of the appeal she held for the public, many newspaper articles were written about her, featuring the white-haired lady growing ferns and California natives at the end of the challenging dirt road in one of the most beautiful settings on earth. By garnering this critical publicity throughout the years, Yerba Buena Nursery helped thrust native plant horticulture into the limelight. Frequently photographed and often interviewed, Gerda was sometimes bemused by what journalists found newsworthy about her. "They always focus on my work boots, or on the fact that I can still push a wheelbarrow," she noted.

She did not fully comprehend how hungry we are for models of people who are capable of sustaining and being sustained by an interest through a lifetime. Growing native plants and ferns, living in her beautiful house, healthy and vigorous at an advanced age, working on the same land for fifty years, Gerda was a novelty in the second half of the twentieth century. Her contributions to native plant horticulture are inseparable from who she was and the way she functioned in the world.

Through the years, her vision grew, aided and abetted by the ideas and skills of the employees she attracted. Because so many different kinds of people with widely differing interests were drawn to work with her, Gerda never needed to advertise for an employee. They seemed to magically appear when needed. She had a special gift for discerning and fostering special abilities. If someone was interested in manzanita cultivars, then the nursery grew manzanita cultivars. If someone was interested in seed propagation, or in native grasses, or in ceanothus, they were allowed to specialize. If somebody had an aptitude for irrigation systems, the nursery sprouted irrigation systems.

Gerda did not introduce dozens of cultivars to the trade, though those who worked with her have. (Gerda walking over to the *Arctostaphylos* section of the nursery and sighing in exasperation, "All these cultivars," is fresh in my mind. Indeed, many of those cultivars are no longer with us.) She did not supervise the planting of many native plant gardens, though those who worked with her have. She did not write about native plants or engage in ecological restoration, though those who worked with her have. Gerda herself and the slow organic growth over the last forty years of one of the most comprehensive native plant nurseries in California comprise her contributions to native plant horticulture. Gerda embodied a quality that some call "generativity," or, as her oral historian, Suzanne Riess, commented, "Gerda is someone who has set into motion an inordinate number of good things."

GERDA'S STORY

In the Quonset hut at Yerba Buena Nursery is a massive cement table where thousands of plants have been potted up in the forty years of the nursery's existence. Built by Hildegard Sander Jackson thirty-five years ago and still used for the same purpose today, that table seems immortal. Nursery workers have come and gone, buildings have been demolished and others built, yet the camaraderie peculiar to the potting table still flows across its rough surface.

Over this table, many of us first heard Gerda's stories about growing up on a self-sufficient estate called Travenort, near Hamburg, Germany. Between twenty to thirty families lived on the estate, which included a hundred dairy cows, pigs, horses, fields of grain and beets, vegetables and flowers, hothouses and greenhouses, and a large house staff, including a butler, chauffeur, and nanny, as well as many cooks, maids, gardeners, and farmhands.

Even as a child in Travenort, Gerda demonstrated a strong impulse toward egalitarianism. Though servants in her home referred to her parents in the third person—"Does her gracious lady wish something?"—Gerda formed friendships on the estate that did not meet her family's approval; she befriended the estate manager's daughter and loved working on the land with the estate's head gardener. After reading Tolstoy's *Anna Karenina* (in German, French, and English) she demanded to be shown how to scythe hay, and he complied. (Once, in her seventies, she demonstrated how to use a scythe, mowing down weeds in the demonstration garden with an effective, even-handed swing that none of us could match.) Her affection for her adopted land of the United States was based in part on the resonance she felt with its democratic principles, however imperfectly applied. Before Yerba Buena Nursery absorbed her energies, civil rights was her life focus.

Income from Hawaiian sugar plantations—derived partly from the disruption of Hawaii's native flora—funded the family's lifestyle in Germany, but in later years Gerda redirected it to the furtherance of California native plant horticulture through Yerba Buena Nursery. It is my impression that most of Gerda's extra resources went to improving the nursery, or to bailing it out in slow times.

Her own life in the twenty years I knew her leaned toward the ascetic. Since Gerda was a Quaker, in some regards she exemplified the simple life. One blue suit, one green suit. One red checked wool jacket for winter. One reliable bread, pie, and cake recipe. Spare and pared-down bedrooms, skimpy towels. If the nursery turned a profit, which was not always the case, most of the money went back into the physical plant.

Frau Isenberg would walk slowly through the rose beds of her formal gardens, snipping off dead blossoms and letting them fall to the ground. A gardener walked behind her to pick them up and carry them to a rubbish pile. When I first began working at the nursery, I watched Gerda at seventy-five and her friend Marielis Forster, not much younger, working late into a December afternoon, laying down gravel on nursery pathways. As the day darkened, these petite ladies heaved shovelful after shovelful of gravel out of the wheelbarrow. At a critical juncture, Gerda had received training that sent her in a very different direction from her mother.

In the normal course of events, Gerda would have been sent to a finishing school in Switzerland. Because of economic conditions in the early part of the century, followed by the chaos of World War I, she was sent instead to a garden school. From her stories, I got the sense of an isolated young girl, not allowed to play with the children of the estate, who longed for something real to do and companions to do it with. The garden school was heaven for Gerda. Talking about those days, the phrase she always used was, "We had a ball."

Being with her peers all day long, working with her hands, learning skills that she would practice for the next sixty years, swimming in ponds, taking walks and playing the guitar, this solitary young girl blossomed. Though they worked hard and ate food so bad that she remarked on it sixty years later, she experienced this schooling as a time of freedom and fun. Her sociable, essentially democratic nature came into ascendance. She fondly remembered the Russian peasant conscripted to work with them on the farm. "Funf und zwanzig," he would scowl fearfully at the girls if they left a tool out or made some other mistake. "Funf und zwanzig," he would mutter ("five and twenty lashes").

At age twenty-one, she was invited to visit relatives in Kauai, the seat of the family fortunes. Here Gerda went surfing, learned the hula, fished and met her cousin, Ruolf Isenberg, who eventually proposed marriage. When I tried to get her to say something about the relationship, such as "Were you in love?" she said, "I suppose so."

When she returned to Germany, he persisted in his suit. As Gerda put it, "My relatives wore me down," and finally she acquiesced. "I didn't have anything else planned," she said. This may have been the last time Gerda didn't have anything else planned. Her parents demanded that Isenberg show evidence of a profession before they married, so he bought a 75-acre strawberry farm in Carmel, which became their first home.

In 1928 their first child, a daughter named after Gerda, was born. Returning from the hospital, Gerda found that in her absence Ruolf had sold the strawberry farm. They moved to Los Altos, in Santa Clara County. As she said of her life then, "I was always trying to do something positive, to create something to work at, to find a project." Always a voracious reader, in 1931 she created an outlet for her love of literature by opening a bookstore. Located in San Francisco, the European Bookstore was a center for émigrés from many countries.

During World War II she became involved with the American Friends Service Committee, which became a lifelong connection. Working with Josephine Duveneck of Hidden Villa Ranch, she aided Jewish refugees from Europe and Japanese American internees, writing eloquently about the conditions of the latter. Gerda's efforts earned her the lasting gratitude of many Japanese families and a warning from the FBI. At one point, a swastika was burned on her lawn. After the war she worked with the Fair Play Council, tackling issues of racism, low-income housing, and farmworkers' conditions. In 1950 Gerda ran, unsuccessfully, for the California State Assembly.

Her husband bought the three-thousand-acre cattle ranch that would later provide the setting for Yerba Buena Nursery. They moved there to live in a primitive house with, as Gerda matter-of-factly put it, "five adults and six children, four of our own, a niece of my husband's, and a refugee boy from Germany, plus the music teacher, who later married my husband."

The ranch kept Ruolf busy for a while, building roads, digging three ponds, fixing old barns, and herding cattle. When he decided it was time to cash out and move on, Gerda balked. Both of them held title to the ranch, and Gerda refused to sell. With his next wife, the aforementioned music teacher, Ruolf Isenberg would move seven more times. Gerda said, "I saw that coming and said 'No more.'"

So she kept the land on which the nursery was built and lived there for over fifty years. Gerda remarked, "I was raised to start something and stick with it." She chose commitment to a place over the seeking of new challenges. Gerda's connection to the ranch and the sense she gave of a person deeply in tune with her surroundings became part of the attraction that drew people to her and to the nursery.

After the divorce, Gerda looked around for a source of income. With the

FIGURE 5. Gerda Isenberg's bookplate, by her daughter, Ami Jaqua.

encouragement of her longtime friend Charles Burr, she began to grow ferns, which he thought would be popular, and native plants, regarding which he said, "What do you want to grow those weeds for?" I tried to elicit from her some description of how she first got the idea of putting containers of *Mimulus aurantiacus* in her station wagon and driving to all the nurseries on El Camino Real to sell them. Interested in all plants, she had been exposed to natives like sticky monkeyflower while riding the range for their cattle ranch. "They were what was here," she said. "I did it because I thought these plants had a horticultural value. And why would customers come all the way up here to buy the same plants they could buy down there?"

Her first employees were two German Jewish refugees, Gertrud Aronstein and Helmut Schneider. Her first employee of the group I think of as "the sixties group" was Judith Skinner. Gerda had an interesting relationship to the decades of the 1960s and 1970s. For many idealistic young people, the nursery provided a place where good work could be pursued and knowledge gained in compatible circumstances, a situation where integrity reigned. Employees were never asked to fabricate or even to exaggerate the claims about native plants. On her side, Gerda watched the excesses and experimentation of those times with remarkable tolerance, considering her strong inclination toward the golden mean.

As the years passed and the nursery grew, interest in native plants became more commonplace, leaving far behind the days James Roof described, when only ten people in the Bay Area shared an interest in natives. A procession of workers benefited and were benefited by Yerba Buena Nursery. Different groups formed and reformed around the lunch table, different combinations of personalities clashed or harmonized over the potting table. Improvements were made: a new propagation house, a new fern house, a new office, a new watering system. The demonstration garden was fenced, paths made, thousands of native plants grown from cuttings and seed. Droughts came and went, each one stimulating a flurry of interest in native plants and usually a newspaper article about Yerba Buena Nursery. The nursery faithfully supplied items for the monthly raffles and yearly plant sales sponsored by the Santa Clara Valley Chapter of the California Native Plant Society. Countless Bay Area home owners found plants for their gardens at Yerba Buena, inspired and informed by its demonstration garden and its knowledgeable, dedicated employees. In 1980 Gerda became a Fellow of CNPS.

The Yerba Buena Nursery Demonstration Garden, an important part of Gerda's dream, became a place where the beauty, growth habits, and requirements of California native plants in the landscape could be seen. Now called the Gerda Isenberg Native Plant Garden, it still receives continued attention and is a valued resource for Bay Area gardeners.

The nursery was a place where a rare kind of information could be acquired and invaluable practical experience gained, surrounded by natural beauty. Part of Gerda's genius lay in her ability to delegate authority, letting people make their own mistakes, encouraging them to experiment. When I began doing seed work at Yerba Buena, I knew only enough to know how little I knew. I would rush into the library, grab *The Woody Plant Seed Manual,* for which I've ever since had a soft spot, speedily look up the species under discussion, then stroll back out to tell Gerda what pregermination treatments were required. Many other employees reported similar stories, stories of being pushed beyond their knowledge by what seemed like Gerda's ignorance of their ignorance, but which may have been something else, a brilliant strategy on Gerda's part to bring out the best of which we were capable.

From Gerda, I learned a work ethic, one in which the needs of the plants dictated activities. She taught by example how to proceed through a day at a steady pace, moving from task to task. Gerda knew that life was maintenance; from her I absorbed this knowledge. Her love of routine was steadying, a structure around which a person could stay productive over the long run. Gerda had a very long run indeed, which she managed with grace and a consistently outward focus. As a Quaker, her inner spiritual life was expressed through silence.

In my experience, there was not a mean or spiteful bone in Gerda's body. I believe that Gerda had the best manners of anyone I ever met. Deep courtesy was extended to both customers and employees. Customers from far away were invited to share coffee and cake in the house. Employees were included in Christmas wreath-making activities and many family social events.

Try as I might, I never could elicit from Gerda her early motivations for growing native plants. It was as though the reasons went too deep for words. I suspect that she unconsciously extended the courtesy that was part of her nature to her adopted land and the plants of her place. Arising from this innate sense of propriety and respect was a love of and commitment to her chosen field that lasted her lifetime. Clear as a bell in my memory are specific moments of Gerda's excitement over new plants and successful growing projects.

Every field trip, every visit to botanic gardens produced seeds and cuttings. We could grow anything that interested us, using the demonstration garden to see if it had horticultural value. It has always been true that you can get things at Yerba Buena that nobody else carries, including the obscure, the difficult to grow, and the rare. Lack of salability was never a reason not to grow something.

While the nursery grew, Gerda met her share of life's difficulties and tragedies. A son and a granddaughter met untimely, tragic deaths. Many treasured friends died. Employees moved on. Gerda even became the victim of a calculated financial swindle. I watched her handle all these events. Intensely skeptical of therapy, she told me, "When I get depressed, I just work." While we watered plants in the fern house, she told me that her granddaughter had died. That was a day when much hard work took place.

Underlying all those years was a powerful theme, Gerda's dream that Yerba Buena Nursery would become a garden school, reminiscent of the life-changing situation she herself had encountered at age sixteen. She hoped to establish a nonprofit foundation, to continue when she died. A board of directors was formed, lawyers hired, the efforts of many hardworking people combined. Not infrequently, while eating in the dining room, Gerda would begin to imagine how it would be when she was gone and her house would be turned into a place for the interns to stay. She relished the images of them eating where we were eating, sleeping where she slept. I used to imagine those interns of the future, chatting away over lunch, fingernails indelibly dirty, learning a trade, "having a ball."

Ultimately these plans did not come to fruition. In 1994 Gerda said, "I've given up." I watched her accept defeat at ninety-four and swim with it. Though never officially a garden school, Yerba Buena Nursery did attract and train interns from all over the country as well as some from other countries. Today, under Kathy Crane's management, interns still find opportunities to work and learn at the nursery.

Visiting Gerda during her last years, I came to regard my overnight stays with her in the light of a retreat, during which I, a completely unroutinized person, could rest in the structure that was second nature to Gerda. In the morning, I would be given an egg in an egg cup and toast with homemade jam. Then we would work till lunchtime, come into the house, wash our hands. Gerda's hands were a study in what years of immersion in soil and plants could do. She sometimes used Clorox to get the deeply embedded dirt out of the deep fissures in her well-used hands, never protected by gloves.

After lunch came Gerda's inviolate twenty-minute nap, followed by four more hours of work, then dinner, with one glass of wine apiece, while we talked of many things, topics determined by Gerda. Though her deafness was considerable, somehow I always felt heard. She talked about the current employees, holding each in her mind, weighing their talents and foibles, each name held lingeringly in her clear regard. She valued the attributes of each, her only criticism concerning departure from the golden mean. "It's too much," she would say, of some youthful love, or aberration, some out-of-balance predilection.

During one visit, when Gerda was ninety-four, we made bread after supper. Witnessing the slowed-down dance of bread making, I learned that economy of movement was the goal. The bread bowl was removed from the same cabinet where it had been kept for fifty years and placed on the table in front of her. Using the table to lean on, Gerda reached, without looking, into the flour bin behind her, grabbed a handful of flour, and flung it into the bowl on the table before her. Several more handfuls followed, till the right amount of flour was in the bowl. Water was added, the dough was kneaded. I learned that taking it

FIGURE 6. Gerda Isenberg at age eighty. Photo courtesy of Ami Jaqua.

slowly, sticking to the same recipe, a baker in her midnineties could turn out good bread.

A predictably humiliating round of Scrabble usually followed dinner, during which I would be soundly trounced, though English was my native language and not hers. A weekly game with her neighbors kept her in good form. Then reading and early bed, the only sound the rushing of the creek through Gerda's garden, the creek that was the source of the nursery's water, making possible its existence.

In 1995 it became clear that Gerda could no longer live alone. She moved to an adjoining parcel of land, where her daughter and son-in-law, Ami and Richard Jaqua, had built a house. Soon afterward the nursery was sold to the intrepid Kathy Crane, who, with no horticultural background, became the fastest learner in the history of fast learners at Yerba Buena Nursery. Kathy made the decision to keep the nursery totally focused on natives, and so it is today.

Kathy visited Gerda weekly, keeping her abreast of changes and decisions, and received in return Gerda's whole-hearted support. Gerda told me that she never offered advice, waiting till it was solicited. With the grace and courtesy that characterized her whole life, Gerda enjoyed the innovations and changes that came to the nursery and was a good customer for high tea at the new Yerba Buena Nursery Tea Terrace. Very near the end of her life she spent an afternoon at the nursery, talking to yet one more reporter, still publicizing the nursery from her wheelchair. When she talked to me about the nursery, I saw how invested she still was, at ninety-five, in the basic act of supplying California's native plants to those who wanted them. It was her part in the chain of life's work, and she still felt its tug, though I was startled to look at her hands, no longer occupied at that work, fingernails eerily clean.

Not long before she died, I had the opportunity to be Gerda's amanuensis. She dictated a letter that was long overdue and weighing on her mind. We wrote and rewrote that letter, until it was perfect in its simplicity, straightforwardness, and logical progression. "I'm very sorry," she wrote to Everett Butz of Wapumne Nursery, "that I didn't respond earlier to your letter of May 3, 1995. I was having hallucinations." This matter-of-fact allusion to a daunting experience is typical of Gerda's attitude toward the whole business of aging. "What do you expect, if you live to be ninety, or ninety-four, or ninety-five?" she would say.

On June 11, 1996, the day Gerda died, Betsy Clebsch and Judith Skinner, friends and employees of Gerda's, picked up from the publishers Gerda's last project, a book called *Yerba Buena Memories.* It is a compilation of letters elicited from thirty-nine Yerba Buena employees and interns, describing their memories of their time, however brief or long, at Yerba Buena Nursery. These letters, written by a diverse group of people, have several things in common. Most of them remember the pleasures of lunchtime at the nursery on Gerda's patio. All of them mention the beauty of the natural surroundings of the nursery. Last, they describe their sense of good fortune in crossing paths with Gerda Isenberg and Yerba Buena Nursery.

In her early years with the nursery, Gerda met Lester Rowntree several times, and she treasured each of those meetings. When Lester was one hundred years old, Gerda decided that she needed to see her just one more time. So she made the long drive to Carmel. When she arrived at Lester's house, she was told that Lester was no longer able to have visitors. Many times I heard the story of Gerda's disappointment that Lester was no longer available to her. I didn't fully understand what she was saying. Now that Gerda has been gone for ten years, I do.

Now that Gerda's unique brand of incisive understatement, her appreciation of our efforts, and the rare sweetness distilled from a lifetime's generous experience, are no longer available to those who cherished her, we can recount the life lessons that rose from knowing her. Among them, I include the way she was cared for by her daughter and son-in-law when she became infirm, how she handled the challenges of enforced immobility, and what it is like to mourn the death of a person who has lived their humanity as fully as can be imagined, whose life seemed as beautiful as it is possible for a human life to be. There are no regrets here.

Gerda visited me once at Larner Seeds in West Marin. I showed her my small seed company, and my newly developed demonstration garden. At the entrance, I had planted one of Yerba Buena Nursery's introductions, *Ceanothus* 'Gerda Isenberg'. Some years earlier, over her protests, we at the nursery had insisted on giving her name to this selected form, which has a particularly beautiful fountain-like form. The specimen growing in my demonstration garden was still small, so I identified it for her.

"It's *Ceanothus* 'Gerda Isenberg'," I said, and I bowed to her, and Gerda bowed back.

15 Forest Gardens
The Lessons of Coarse Woody Debris

Debris: The remains of anything broken down or destroyed.
Webster's Dictionary of the English Language

It is easier to put a person on the moon than it is to manage one acre of forest.
Chris Maser, *The Redesigned Forest,* 1988

I am the caretaker and student of eighty acres of land in Mendocino County, north and inland from my north central coast garden. Gardening in a restoration manner on this land, which contains both forests, with a continuous canopy, and woodlands, with a discontinuous overstory, is a completely new challenge. As some of us move back into the trees (and according to forester Ray Raphael, private owners now control 57 percent of the forested land in the United States), a significant portion of wooded and forested second-, third-, and fourth-growth timber has come into the hands of people whose inclination is toward protection. What might that look like? Here I wander around like a child, searching for clues about how best to interact with the woods, slowly turning what I learn into ideas that I can present to clients who are also newly responsible for wooded or forested land.

On a suburban street near downtown Saratoga, a carefully landscaped yard contains one surprising feature, a Douglas fir cut off ten feet above the ground and left in place, branchless and dead. Tacked to the snag is a laminated note, explaining to the concerned neighbor or curious pedestrian the reasons for the presence of this unusual garden ornament. This standing dead tree is one of

With thanks to Dr. Lucinda Johnson, Department of Biology, University of Minnesota, and in memory of Martina Johnson-Kent. Nothing is lost, nothing completely gone.

the three kinds of *coarse woody debris,* defined by forest biologists as any woody material no longer living that was once at least eight inches in diameter at breast height. This snag might become a home for pileated woodpeckers, a granary for acorn woodpeckers, and a source of food for beetles, fungi, and bacteria. As it decays, it adds to the biodiversity of this garden and the neighborhood.

On another landowner's sloping wooded lot in northern California, we find the second and third kinds of coarse woody debris, dead trees on the ground and dead trees in the water. Here, in the redwoods, dead trees and dead tree parts are treated like treasured resources, considered an essential, useful, and interesting element of this landscape. Most are left to lie where they fell, but some smaller and more easily moved dead trees are used to lessen potential erosion problems from road cuts and construction damage. Some are arranged parallel to the slope. Others, with complex branching structures still intact, are set into the hillsides with trunks perpendicular to the slope, a technique called *vegetative riprap.* Major limbs hug the hillside, providing small niches where soil can accumulate and plants begin to grow. Branches and whole trees are placed with their base snugged into the base of the slope, smaller branches and twigs heading upward. The niches formed by these numerous crotches accumulate soil and make good nurseries for seed and transplants. Compared to other methods used to stabilize minor road cuts and stream banks, such as cement sacks, cement blocks, or imported rock, vegetative riprap can be an unobtrusive, natural-looking way to hold slopes until plants become established.

The gardeners at these sites are enlightened practitioners of the restoration arts, including in their yards a place for coarse woody debris. The presence of significant coarse woody debris is a crucial element in old-growth forest. In old-growth Douglas fir forests, 25 percent of the total biomass consists of coarse woody debris, standing or downed logs in one or another of what biologists have separated into the five stages of decay before final incorporation into the forest floor. The complex and varied accumulation of different kinds of woody debris in the forest is the keystone of its biodiversity.

With the demise of our old-growth forests, some of the important ecological functions of coarse woody debris are no longer being performed. All those who live near woods have the opportunity to become a knowledgeable fan club for standing dead trees, dead trees on the ground, and in some fortunate situations, dead trees in water. For gardeners with wooded land, awareness of coarse woody debris is an enriching piece of the puzzle of their terrain.

PLATE 1. Traditional seed beater and jars of seeds after the summer and fall harvest. Photo © Eirik Johnson.

PLATE 2. Male quail with ruby chalice clarkia and rosy buckwheat at Golden Gate Park, San Francisco. Photo © Alan S. Hopkins.

PLATE 3. Coyote bush as the keynote plant framing the Hermit Hut, a restoration garden feature. Quail take shelter under this stand of coyote bush during rainstorms. Photo © Peter G. Smith.

PLATE 4. Western scrub-jay pausing on coyote bush.
Photo by Rich Stallcup/PRBO Conservation Science.

PLATE 5. Vine maple, one of the showiest native shrubs for fall color, shown here in October in a northern California garden. Photo © Saxon Holt/PhotoBotanic.

PLATE 6. Indian lettuce, winter's green tonic, beginning to bloom. Photo © Peter G. Smith.

PLATE 7. Section of decomposing log invaded by roots of the hemlock tree, in a mycorrhizal association with mushroom *Russula emetica.* Photo © Gary Braasch.

PLATE 8. Spring wildflower field at Tejon Ranch near Gorman. Photo © Sonja Wilcomer.

PLATE 9. Tansy leaf phacelia, a fragrant and vigorous annual beloved by insects. Photo © Peter G. Smith.

PLATE 10. Meadowfoam, a vernal pool wildflower that thrives in the Larner Seeds garden. Photo © Peter G. Smith.

PLATE 11. *Limnanthes douglasii* subsp. *nivea* in Mendocino county, with female native bee *Panurginus occidentalis.* Two of the "small things" that go together and can be promoted in the backyard restoration garden. Photo © Peter G. Smith.

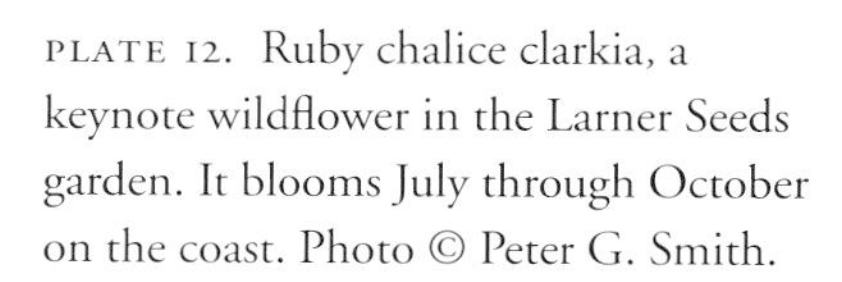

PLATE 12. Ruby chalice clarkia, a keynote wildflower in the Larner Seeds garden. It blooms July through October on the coast. Photo © Peter G. Smith.

PLATE 13. Red ribbons clarkia, an annual wildflower that helps to give the garden a local flavor. Photo © Lottie Jenvey.

PLATE 14. June bloom of perennial coast lotus, one of the forbs that make good container subjects. Photo © Peter G. Smith.

PLATE 15. *Linanthus* sp. from Mendocino County growing in a container, where it has reseeded itself for three years. Photo © Peter G. Smith.

PLATE 16. Elegant madia in August at Larner Seeds, brightening a foggy day. Photo © Alan S. Hopkins.

PLATE 17. Purple flowering shrub lupine and blue blossom, two shrubs that are appropriate in size for the small garden. Photo © Peter G. Smith.

PLATE 18. Chair next to a pot of dormant California fescue—a good place to rest in the fifth season. Photo © Meleta Kardos.

PLATE 19. Tule reflected in the pond at Larner Seeds, flanked by coyote bush and blue blossom. Photo © Eirik Johnson.

PLATE 20. Close-up of twinberry, a shrub that tolerates wet feet and provides berries for migrating birds. Photo © Peter G. Smith.

A GOOD TIME TO ENJOY COARSE WOODY DEBRIS

In the middle of a rainy winter, when many parts of California are receiving twice their normal rainfall, when hillsides are sliding, creeks overflowing, and floodplains asserting their true nature, I go to my wooded field station in Mendocino County to check on the progress of our coarse woody debris. On the drive north, I anticipate examining sodden logs on the ground, with all manner of relationships to each other, to the soil, and to the invertebrates, amphibians (such as the Pacific giant salamander and the western red-backed salamander), fungi, and mammals that inhabit them. I can hardly wait to see what the season has brought.

Before I even get out of the car, I can see that a whole new series of coarse woody debris events has transpired in my absence, making an exciting contribution to the forest floor. All around me, both new and previously downed logs in a dizzying variety of situations create a wide range of homes for a wide range of creatures. What was heavy rain in my coastal garden eighty miles south was a once-in-fifty-years snowstorm here, and the results are obvious. While not a *debris torrent,* in which an entire sodden hillside gives way, trees and all, it is still a significant event. I see before me a veritable garden of decay scenarios, requiring no input from me, just admiration.

For the previous five years I've watched twenty rapidly growing Douglas fir saplings surround a sun-loving Pacific madrone. Shooting up toward the light, they rapidly shaded out first the madrone's lower limbs and then the midlevel branches, finally overtopping it. As the years passed, the madrone's sleek, sinuous limbs curved and contorted ingeniously, as only *Arbutus menziesii* can do, searching for the sun.

This year the madrone, made leggy and unwieldy from too much shade and competition, succumbed to an early winter windstorm, which ripped it out by the roots. Falling downhill, it crashed into a mature Douglas fir, which reeled from the impact and snapped off fifteen feet above the ground. Shelf fungus had already weakened the fir, increasing its vulnerability to this collision. Because the Douglas fir was large, over two feet in diameter, the remaining snag is big enough to house a black bear. The disease called heart rot is the architect of a black bear's dream home, carving out a cozy den.

The crown of the snapped-off Douglas fir then fell onto a large tanoak below it, and that tree too snapped off and fell straight down the hillside, leaving one more snag. Its fallen branches cover the ground like a catcher's mitt, forming a leafy bower, under whose privacy voles and other mammals take shelter.

Taking inventory, we now have one downed madrone, death caused by wind throw, lying with its trunk mostly in contact with the hillside; one downed Douglas fir, death caused by insect damage, disease, and the consequences of wind throw, its top partially dug into the hillside; and one downed tanoak, whose trunk arches off the ground. We have two newly formed snags, one suitable for an acorn woodpecker home or granary, and the other large enough for a black bear or a pileated woodpecker. We have one newly exposed root-ball (the madrone's), uncovering previously hidden mineral soil, making a good nursery for certain tree and shrub seedlings. This event has created possibilities that will extend far into the future, but it provides only a small sample of the many possible scenarios and variables affecting the mode and timing of incorporation of trees into the forest floor.

TIME IN A WOODLAND GARDEN

Deep in the forest, I sit on the stump and enjoy this warm, sunny, late winter day. I luxuriate in the sense of nothing to do, nothing to fix. As the sun heats up the forest, phenols from the conifers that surround me pour into my nostrils. Birdsong from unknown birds clears my brain, and I will not try to name them.

I ponder my coastal garden far away. That focus of much effort, that one acre of intensely restored, intensely managed land, lies eighty miles south and right on the coast. For twenty years it has received a certain kind of concerted attention. Though these woods in which I now sit have been logged twice, burned once, and grazed for years, they do not present me with the immediate and compelling call to action I experienced when I first saw my land on the coast. That land, reflecting little of the native flora, made urgent demands upon me. Living in and restoring coastal scrub, my coastal garden talked to me in various ways, but not in terms of coarse woody debris. Coastal scrub components are relatively dainty, the return of their cellulose to the land comparatively insignificant. I come to these woods for a different experience, different calls to action, different kinds of peace.

It is certainly no pristine forest. The size of the stumps remaining from the last logging attest to the fact that my musings take place in a very changed environment. There were giants here. But the cast of characters, as far as I can tell, is still fairly intact. There are openings in the woods where fragrant buckbrush

(Ceanothus integerrimus) sends its enticing aroma into the open window of our car as we drive slowly by. If the window is open, a branch may swipe us, sending a shower of tiny white spent blossoms into the car, and we don't mind at all this lovely shedding.

So what is my role here? I have just begun to notice the irregularities in the forest floor in our wooded field station. To avoid tripping and stumbling over what seemed to be but was not a uniform surface, I learn to look down as much as up. I note that these lumps are remotely tree-shaped and begin to realize they once were trees. This unevenness of the forest floor, called *microtopography,* is primarily the result of the addition of coarse woody debris, which takes place in a stunning variety of scenarios.

WHY A TREE FALLS

Trees fall for different reasons, and each reason creates its own set of consequences. The three possible causes are mechanical factors, insect damage, and disease. Mechanical, or physical, factors include wind throw, the impact of another tree falling on it, ice, snow, logging, fire, and lightning. Nothing is trivial; each circumstance creates the conditions for specific events. If fire was involved, was it burned before or after returning to the forest floor? Was the tree alive or dead when it hit the ground—if alive, what weakened it; if dead, what killed it?

In our story, the protective exterior layer of each tree will be breached differently. The Douglas fir's outer bark was already open to invasion. The tanoak split from the impact of the fall. The madrone will experience the ministrations of beetles, who will bore holes through the outer bark to the tender, living cambium beneath. After the beetles, a staggering variety of wood-eating insects, some of which specialize in phloem, others in xylem, will eat, compress, lay eggs in, make chambers in, and prepare the tree for subsequent inhabitants.

The species of the tree may determine some aspects of the decay drama. Different species attract different invertebrates. Some wood-eating insects will eat only certain species; others are less fussy. Some tree species are attacked while living and come down to the forest floor with their inhabitants already present. Though many wood-eating insects have been identified and named, the life histories of many others are still a mystery, and new species are continually being discovered.

GOING TO GROUND

The way the tree lands is another major inconstant that has a great deal to do with determining the speed with which the tree will move through the five decay classes. Whether the tree falls parallel to or perpendicular to a slope, with all parts touching or with only some contact points touching the ground, is of significance. It may have ripped out by the roots so that mineral soil is exposed, making an opportunity for a whole different suite of animals and plants, or it may be arching over the ground, in touch at either end with the earth. In that case, the distance from the ground is a determining factor in its future story. Total contact with the forest floor, the fifth class of the five-class system of decay devised by Chris Maser and others, is its ultimate destination. How long it takes to get there and what happens along the way are the distinctive elements of its particular destiny. As the tree progresses through the classes, it is first elevated on support points, the remains of branches or twigs. These sag slightly in class 2 and are sagging near to the ground in class 3. In class 4, all of the tree is on the ground. Finally, in class 5, the tree's texture grows soft and powdery, almost indistinguishable from the forest floor.

These woods here, after this powerful winter storm, are rich in class 1, trees newly on the ground, with intact bark and twigs, original shape, original color, trunks elevated on support points, and no invading roots. Class 4 coarse woody debris is also well represented. I am particularly fond of the pale reddish-orange chunks of partially digested Douglas fir sapwood, bark, and heartwood characteristic of class 4. Sometimes I dissect these logs, pulling the chunks apart to use as mulch or for paths, hastening them on their way to class 5. Characteristics of bark, twigs, texture, shape, color of wood, portion of tree on the ground, and part of tree where invading roots are present follow their sequential course.

Also relevant to the scenario is where the tree fell and when. Is it in sun or shadow? Did it fall during the wet season or the dry? The properties of the soil upon which it lands, whether it be mineral, loam, bedrock, or wetland, are of significance: decay rates in wetlands, where moisture is constant, can be low.

A FALLEN TREE'S ESSENCE

The fallen tree's course through time is also determined by age, species, and health when falling. The age of the tree determines what percentage of the tree was sapwood or heartwood, young trees having a higher proportion of sapwood to heartwood. Sapwood and cambium, still full of nutrients and the juice of living matter, attract decomposers that work quickly; hence young trees have less staying power. Heartwood, nonliving, is attended to by a different class of decomposers, with more complex digestive systems. Nutrient cycling in old-growth forests has a different pace and a different cast of characters than in young or mature forests. Looking at coarse woody debris, we learn deep reasons to preserve what old-growth forest still exists.

> Habitats provided by the death of young trees are short-lived and rapidly changing. In contrast, the less frequent, more irregular mortality of large trees in old forest is analogous to slow release fertilization. The lasting quality of large fallen trees creates stable habitats in which large woody debris accumulates. . . . Large fallen trees in such an area often contact each other physically, creating external habitats of intense biological activity.

We will not see again the giants of old, who, when they fell, were still taller prone than a tall man standing. Their massive component of lifeless heartwood was an energy sink, a place for nutrients and moisture to be held and slowly released over a long period of time. Their tenure on the forest floor could be as long as four hundred to five hundred years. The real consequences of the lack of coarse woody debris of substantial size in our forests won't be known for generations.

THE POETS OF COARSE WOODY DEBRIS

When I first read about coarse woody debris and discovered the classic "decay literature" of the 1960s and 1970s, I felt like Keats first looking into Chapman's Homer and compared myself to Cortez, "silent upon a peak in Darien." A whole world opened up, and I was forever changed. Some of the early works, like *From the Forest to the Sea, The Seen and Unseen World of the Fallen Tree, The Redesigned Forest,* and *Forest Primeval,* are classics, elegantly written, whose every sentence is so punched up with material for metaphor that I read and re-read, underline and outline. Authors like Jerry Franklin, Chris Maser, and James Trappe should be well known and revered, their fame not limited to the audiences at their own conferences. They are masters of the law of consequences, the law that says that every event sets off a series of other events, extending far into the future.

The biologists studying coarse woody debris demonstrate an ability to take the long view. Experiments with a time frame of two and a half centuries can be part of their field of study. Their observations require the patience to slow down, the fortitude to attend to the details of the specific dramas that attend each stage of decay, and the insight to give words to what we may not know we see. What they present frequently seems like common sense, but of the most sublime kind. They remind me of those Buddhist monks who meditate on corpses in charnel houses and graveyards. Attentive to each tiny consequence, they have strong stomachs for rot, feces, decomposers, parasites, insect vomit, and other matters relevant to nutrient cycling.

They teach that everything has significance. Each event contains a specific and never to be duplicated set of circumstances and consequences. If the tanoak had died from disease, dead when it hit the ground rather than killed by wind throw, different characters would participate in its decay drama. If the windstorm had hit in early autumn, no fawn would have found a nursery under its sheltering boughs. If the madrone had fallen sidewise, landing along rather than pointing down the hillside, soil would accumulate above its trunk, creatures that like such situations would find it, and decay would proceed at a different pace, with different entities acting out their nature. Decay dramas present an array of colorful characters, an orderly procession of actors, each setting the stage for the ones that follow. The details of this progression are in every case unique.

FOREST PRACTICES AND COARSE WOODY DEBRIS

The impetus for the study of coarse woody debris came from timber harvest practices in the 1950s and 1960s. At that time, and into the present day as well, in spite of this body of compelling research, all cellulose was considered fair game. Dead trees on the forest floor could be harvested with impunity. In the "managed" forests of today, where clear-cutting takes place, coarse woody debris is either removed entirely, its quantities extremely diminished, or its quality is rendered uniform, in same-age monocultures.

Research into forest systems gave a value to debris on the forest floor beyond an opportunity to harvest cellulose, a resource that would otherwise be "wasted." Studies explained that old-growth trees decay more slowly than young trees because through the centuries they have formed proportionately more heartwood, which, inert, decays more slowly than the living parts of the tree, resulting in greater stability. Our forests, creeks, and rivers benefit from the slowing down that aged trees create. The metaphors for human society are almost irresistible.

Rain returns nitrogen to the forest, carrying it from the treetops to the ground. And nitrogen returns in another way, discovered by Tom Reimchen, a brilliant Canadian biologist working in the Great Bear Rainforest in Vancouver, BC. Within the temperate rainforests of the Pacific northwest, which includes part of northern California, there is a profound connection between the old-growth forests, the salmon that swim up the rivers to spawn and die, and the bears that drag their carcasses into the woods. By patiently extracting core samples from hundreds of hemlock trees, Reimchen discovered that the trees and the salmon have a common marker, a nitrogen isotope, the rare N15. This element, coming from the middle of the ocean, deposited throughout the forest by bears feeding at night and dragging carcasses into the woods, becomes part of the soil and then part of the trees themselves. The essence of salmon nourishes the woods.

BIOLOGY AS METAPHOR

Old-growth trees are dying for half their life, homes to decay and disease, yet still alive and important. This biological reality helps us understand our own lives, our own cultures. Old-growth trees may be dying for the second four hundred years of their lives, and still contributing even after their death.

My sorrow for the giants that are gone is the kind that sits heavily in the heart. After my investigations into coarse woody debris, I know that the tree giants are still with us, through the stumps and mounds enlivening the forest floor. This information comes not so much by way of consolation, as by way of truth about decay. It takes a long time for things to be gone. In this song of the details of decay, of coarse woody debris, we glimpse our long-term consequences, which are one and the same as our immortality. I still experience something of the essence and consequence of these giants, one hundred years after their departure. As I glimpse their biological meaning, they redefine immortality for me. Death, where is thy sting?

Every little twig of a life element that we drop creates a drama beyond our lives and beyond our imagining. Through the years our influence becomes harder and harder to discern, our effect less something to trip over and more part of a rich undistinguishing medium, but we are never not there, never completely gone.

Looking into the lives of the three women out of many who have helped us to where we are today in the field of native plant horticulture, I am repeatedly struck by the intertwining connections among them. Edith Van Allen Murphey kept in her files an article by Lester Rowntree and was heartened in ways we cannot imagine by her words and her example, though they never met. Gerda Isenberg had in her library Edith Van Allen Murphey's book and visited Lester when she could. Lester bought plants from Gerda, who advocated for the reprinting of Lester's book, which has now happened. Gerda told me a story Lester told her, which I repeated to my employees. Who knows who they will repeat it to? Never not there, never completely gone.

In life we learn things that seem startling at first, then we incorporate them so rapidly and thoroughly into our changed worldview that the moment of first knowing is lost. I feel that I have always been a devotee of the engaging ways of coarse woody debris, noting and enjoying its presence in the form of standing dead trees, dead trees on the ground, and dead trees in the creek. This information has become part of the answer to the question I ask when visiting our forest field station: "What do these woods really want?"

What can I do for this forest, given its own individual history of logging and grazing? Besides thinning, besides vigilant weed control, besides forestalling erosion with vegetative riprap, besides preserving snags, I glimpse what the forest really wants, which to the limited extent of my lifetime is within my powers, and the powers of some of my clients, to give.

The forest wants back the gift of time.

Spring

16 The Flower Dance in Modern Times

Some day our grand-children are going to press around our rheumatic knees and ask for stories of Old California (the California of our young days). I know what I shall tell mine. I shall tell them of the glories of desert bloom after a wet winter, of California's wild flower fields that were. I shall describe acres of alliums, fields of desert lilies, miles of monolopia, tracts of thistle sage, of evening primroses and sand verbena, and the desert floor massed with a jostling crew of phacelias gilias malacothrix and the other annuals. The youngsters' eyes will bulge (or I hope they will) and they will clamor (as I mean them to) to be shown these marvels.

Lester Rowntree, "Wildflower Sanctuaries," 1934

Meantime what had California been doing with these same wild flowers? Ploughing them up, grazing them up, burning them up, burying them under concrete and asphalt, tearing them out by the roots to gratify a passing whim and then cast aside; until, like her Indians, her wild flora has been desolated and largely driven away from the homes of the people to find refuge too often only in deserts and mountain fastnesses. But while there are seeds there is hope; land is still plentiful enough for a bit of garden with every house; and all who will may yet have at their own doors a little Wild Garden of California.

Wallace Hebberd, *The Wild Gardens of Old California,* 1927

And then it happens. Maybe it's the first wild iris, or the lacy flutterings of white milk maids in the woods. Or a night wind with an utterly surprising balminess. Then we see our first shooting star, that inside-out flower with a wide range throughout the state, covering many different kinds of hillsides on many different kinds of soils, yet so far defying introduction into the garden. The procession begins, all over the state, in combinations and single species, bloom and seed set, bloom and seed set, more different species than any other state in the union, more than any of us can ever see in our lifetimes (though some try).

As an ex–New Englander used to harsh winters, I wonder if we have really earned what is about to be bestowed upon us—three or four months of a succession of annual and perennial wildflowers, still with us, still persisting, still unique to California. I know a way to obviate the guilt and feel that I deserve these pleasures. I join the modern flower dance, which has two parts: first, experiencing and rejoicing in the wild bounty that remains, and second, making a place for native wildflowers in my home garden. Both parts are mutually enhancing; the wild gardens give me endless ideas for combinations and arrangements, as well as transcendent pleasure, and the home garden provides an intimacy and opportunity for understanding that I bring with me back to the wild.

In spring, in years gone by, flower festivals, flower dances, and play with flowers took place throughout California. It happened where the Pomo lived at the edge of the redwoods, where the Chumash lived in the "valley of the flowers" in southern California, with the Wintu on the McCloud River, in the mountainous Miwok country, and in the steamy land of the tules, where subtribes of the Yokuts reaped the bounty of their fertile wetlands.

Some of these vibrant ceremonies involved the adorning of young girls with California's annual wildflowers, eliciting a powerful parallel between the beauty of girls coming into womanhood and the ebullient floral displays of a California spring. Along the American River, the Maidu had a dance in which girls wore flower fillets and pelted spectators with flowers. William Shipley describes the spring festivities of the Maidu of northeastern California:

> In the blossom time of early spring, when the dogwood and the first buttercups were blooming and the wormwood was in green leaf, and the women prepared lots of food—game and fish, acorn bread and soup—and the men got out their leather finery, their eagle feather bustles and their yellowhammer featherhead stalls. For several days, rubbed with wormwood leaves for the fragrance, bedecked with garlands of flowers and foliage, the people sing and dance, rejoicing in the return of clement weather. And, on the last day, they all ran down to the river and threw their garlands in, imploring the rattlesnakes to stay away. "Snake, snake, don't bite us!" they shouted. "Snake, look away into some other land! Just as these flowers float away from us down the river, you, snake, go away from us through all the summer and fall!" Then everyone would run back up and feast on all the great food the women had made ready.

Anna Gayton, an anthropologist who worked with Mono and Yokuts tribes in the early part of the twentieth century, describes the flower play enjoyed by the indigenous peoples who lived around Tulare Lake.

> The magnificent sea of wild-flower blooms which covers the San Joaquin Plains from March to May was greeted with rejoicing by the people dwelling there. During this season young people went out to pick the flowers and to construct crowns of flowers for themselves. There were many songs connected with this practice. . . . One song ran, "Now I am making a crown *(se ma)* of flowers." And there was an expression, "to go flowering." Old men who could not wander out in the field would ask children to make crowns for them too.

Profound merit resides in this notion that a young person might make a flower crown for a person now physically unable to make it to the flower fields yet still desirous of wildflower presence and contact. A scene at Arizona's Picacho Peak in mid-February comes to mind. The spaces between the numerous shaded picnic tables were everywhere thick with annual wildflower bloom, in the plenty that requires careful stepping. Many species were recognizable to a Californian—phacelias, gilias, goldfields, clarkias, and poppies. Every family group seemed to include an older person, who was carefully seated, fed, and helped through the flowers, present to be revived and reminded, there to receive a flower crown.

Today thousands of Californians still feel the call "to go flowering." A recent extraordinary wildflower year (2005) produced articles in major newspapers from Seattle to San Diego and rated spots on local TV stations, ravings on websites, and lengthy, detailed recordings on wildflower hotlines. The towns around Death Valley, Anza-Borrego, and Joshua Tree National Park experienced record occupancy. In some places the crowds were severe, as California's wildflowers received their due. For almost every viewer, or every viewing couple, there was at least one camera. Visitors from other galaxies might understandably conclude that placing small rectangular objects between our eyes and the world is the way we humans worship natural phenomena. I myself decided to experiment with burning these visions into my brain, hoping I would be able to recall them at will. Here are some that remain.

THERE'S SOMETHING ABOUT YELLOW COMPOSITES

Though deep blue desert bluebells *(Phacelia campanularia)* making a rich understory for yucca plants and brittlebush were glorious at Joshua Tree, and though miles of desert evening primrose *(Oenothera californica* subsp. *californica)* were lush and lovely, and the desert lily *(Hesperocallis undulata)* a rare pleasure, the image behind my eyelids before I fall to sleep is usually yellow. The yellow wildflowers, many in the Asteraceae family, familiarly referred to as *composites,* with all their shades of cream, lemon, deep yellow, and gold, push a primeval button. Whether yellow-petaled with white tips, such as tidy-tips *(Layia platyglossa),* desert dandelion *(Malacothrix glabrata),* and meadowfoam (*Limnanthes douglasii,* which is not actually a composite but a member of the meadowfoam family), or yellow with gold centers, such as desert marigold *(Baileya multiradiata),* Bigelow's coreopsis *(Coreopsis bigelovii),* and goldfields *(Lasthenia glabrata),* when growing together in great numbers they create the "yellow days" through which John Muir drifted in his famous description of wildflowers in the Central Valley.

The combination of tidy-tips and goldfields is an old favorite of mine, first experienced at Edgewood Park in San Mateo County, then at the Carrizo Plain in San Luis Obispo County, in succeeding years at the Arena Plains in Merced County, as well as at Joshua Tree National Park in Riverside County, again along Shell Creek in San Luis Obispo County, and always in my own garden in Marin County.

To fully experience yellow composites, walk toward them with the sun at your back, so that you view the flowers, frequently sun followers, full face. Coming at them sideways gives another perspective, slightly less yellow, and slightly less vivid. Sometimes tidy-tips and goldfields grow together in equal proportions; other times one may predominate. Whether islands of tidy-tips are set in a sea of goldfields, or the other way around, the edges where they meet create a powerful visual event.

A good wildflower year is an opportunity to teach color to painters, form to sculptors, and joy to the depressed. California's wildflowers offer profound lessons in aesthetics. "Having your colors done" by California wildflowers, correlating flower colors and combinations with experience, is a way to learn color, hue, tone, and shade. Once a client requested that I eliminate all yellows from her wildflower mix. From the satin of the poppy's golden glow to the steadfast lemon of tarweed, she wanted none of it. I sadly did as she requested, but I have come to think that "flower yellows" enter our brain through the optic nerve in a way that may add years to our lives.

I imagine a long room with floor-to-ceiling windows along one wall. A drape at one end can be unfurled across the room, with the light shining through the cloth. In semiabstract form, the shades of cream to gold, from desert marigold to desert dandelion, are thrown across the fabric, titled "Composites of California." The extravagance of the flower fields, with an emphasis on yellow, is revealed only when the entire drapery is unfurled across the room. Cream cups are there, with the white and yellow blending to the cream of meadowfoam. Drifts of orange-gold poppies are midfield, or far in the distance. In many places, a background of goldfields sets off the paler shades.

Goldfields, the genus *Lasthenia,* of which there are sixteen species in California, are annual wildflowers (except for one perennial, *Lasthenia macrantha*) found in vernal ponds, deserts, alkali flats, offshore rocks and islands, woodlands, and grasslands throughout the state. They are individually modest plants that in quantity make things happen visually. They grow usually two or three flowers to a slender stem, each flower no larger than a dime or a nickel, and, like most composites, they have a simple, daisy-like flower. But goldfields are the glue that unite many a wildflower field. In masses by themselves, they electrify, especially when a breeze sends their slender stems shimmering into motion. I have practically levitated while standing in such a field. In masses with other wildflowers, they make other species stand out more brilliantly or merge more gracefully. A year without a *Lasthenia* experience, whether *L. chrysantha, glabrata,* or *californica,* is a plodding year, a meal with no dessert.

Large quantities are not required for this experience to be memorable. I remember one spare goldfield display in the desert, individual flowers far apart, dotted evenly between the desert shrubs, each single flower, dwarfed at only three inches tall, gleaming against the shining white sand. By way of contrast, on another shimmering hillside in the fertile hills of San Luis Obispo County, every inch of ground was covered in gold or yellow or cream. I stepped from gopher mound to gopher mound, reluctant to flatten a single unique combination. A solid field of ten different variations on gold sends a message to the brain: Wake Up.

"A REAL FASCINATION FOR THE EYE AND TOUCH"

When goldfields begin to fade, the flowers lower on the stem go first, turning from bright to dull gold, petals dropping to the ground. Then you know that the time of goldfields is halfway over. The color en masse begins to provide a more muted background for the midspring flowers, like Chinese houses, globe gilia, and bird's eye gilia. At this stage, when pulled apart, the heads reveal gray or black seeds at the base. Roasted and ground, these seeds made a favored flour, called pinole, for the Cahuilla Indians of southern California. Early ethnobotanist and anthropologist David Prescott Barrows observed wildflower seed preparation among the Cahuilla in 1900.

> Some of these seeds are very beautiful and possess a real fascination for the eye and touch. The seeds of the *Lasthenia glabrata,* called by the Coahuillas *ak-lo-kal,* in mass resemble iron filings, being of a dark color and fine elongated shape. They are prepared by being pounded up into a very fine flour, which is eaten dry.

Many of the wildflowers were food, each with its own special attributes, its unique way of being cleaned and separated from the chaff, differing requirements and preparation, varying amounts of oil expressed when ground. Where we have only wheat, oats, barley, millet, and a few other grains, the native Californians had the seed of dozens of wildflowers, as well as shrubs and trees, which provided complex carbohydrates, proteins, high-quality oil, and fiber. Some were eaten alone; others accompanied acorn mush, manzanita cider, or venison. Each tribe had its own preferences, special dishes, and combinations. The disappearance of the flower fields has meant a loss on many levels—no more flower play for the young, no more flower crowns constructed for old men by helpful children, and the eclipse of a whole technology of food preparation based on seeds, which, though not completely gone, is difficult to discern in the mists of the past. Critical indigenous knowledge was tied tightly with access to and preservation of California's flower fields. For the indigenous peoples, the beauty of the wildflowers was inseparable from their promise of seed food.

THE SEED FOODS OF CALIFORNIA

Once an archaeologist let me hold a jar of herbaceous seeds from a site dating back 1,500 years. I recognized this shiny jet-black seed as one of my favorite wildflowers, red maids *(Calandrinia ciliata),* a sun-loving, magenta-flowering annual, a low-growing colonizer that likes bare ground. It bears the distinction of being the only native wildflower to appear on its own in my garden.

I asked a Western Mono elder, a skilled basket weaver, if she or her mother had ever gathered red maids and mentioned the seeds found in archaeological sites in her area. "No, but my grandmother did," she replied. The grandmother of this woman, now in her eighties, might easily have been born in precontact times, the early 1800s, when technologies of gathering and processing dozens of flower seeds still flourished. By 1903, when Pliny Earle Goddard described the

Hoopa Indians on the Trinity River, the gathering, preparing, and eating of the small seeds may have been largely over or at least beginning to decline. He described the preparation of "the seeds of grasses, certain Compositae, and other plants" but with the following proviso:

> The weeds introduced since the coming of white people have so crowded out and mingled with the native plants used for this purpose that the Hupa do not now attempt to gather the seeds. One woman was found who had a small quantity of seeds gathered many years ago. She prepared these in the manner described.

So ended the original California cuisine that provided a kind of nutrition and a range of culinary experiences of a richness and variety that we can only guess at. Or perhaps it only awaits revival, along with the current renaissance of indigenous languages, dances, and land-use practices, ongoing, gaining strength. It may turn up, parts of it, in our own home grounds as well. As the lessons of coarse woody debris taught us, nothing is ever completely lost.

A GIFT OF OUR CLIMATE

The blooming phenomenon that was the source of this bounty is one of the gifts of our semiarid climate. Only in the summer-dry, winter-wet parts of the world, known as Mediterranean climates, will abundant winter rains soften seed coats and germinate seeds that ripened and dropped during the previous hot, dry summer. Only in spring in these Mediterranean climes does this ecstatic culmination of the winter's wet work appear, a cornucopia of colors, fragrances, and forms.

On the East Coast and in the Midwest, and in some parts of the desert, moisture from the heavens can be expected throughout the year, including summer, and perennial rather than annual wildflowers are in the majority. Even after four or five hundred years of logging, farming, and grazing, Eastern and Midwestern woodlands are still graced in many places with masses of shade-loving perennial wildflowers, such as lady's-slipper, bloodroot, trillium, mayapple, and columbine. On the West Coast and in the Southwest, it's the annuals, charged with the intensity of their need to bloom, get pollinated, and form seed before they die under the summer sun, that dazzle our eyes.

The disappearance of some of these flower fields was the work of one generation only, so that someone born in 1880, such as Nicolus Hanson of Winters,

California, could enjoy the beauty of the Sacramento Valley floral abundances as a child and live to mourn their disappearance as an elder in 1944. Grizzled pioneers, moved to rhapsodies by their beauty, lamented the bouquets that could no longer be gathered by children for mothers when European patterns of land use and the invasion of exotic grasses and weeds spelled their demise. These descriptions of the wildflowers are treasures for which we can be grateful. They include praise poems, such as those by former poet laureate of San Francisco Ina Coolbrith and Wilma Elizabeth McDaniel, poet laureate of Tulare County, prayers, and some mournful pieces, which can be elegies for what once was, as well as tears for the poignancy of the beauty that still remains.

WEEPING OVER WILDFLOWERS

The mother of early settler Thomas Jefferson Mayfield was perhaps one of the first to weep for the joy of viewing wildflowers, while overlooking the San Joaquin Valley in 1873. "As we passed below the hills the whole plain was covered with great patches of rose, yellow, scarlet, orange, and blue. . . . Mother cried with joy and wanted to make a home right there in the midst of it all."

In 1993 Stephen Wall interviewed Juanita Centeno (Chumash), living near Santa Barbara. This elder, now deceased, described what she called the Valley of Flowers as it used to be, tears streaming down her face for the flower fields that were no more.

> Wiping away more tears, Juanita added through sobs, "We had our Flower Festival here last year, in June or July; people didn't come. Why should they come? In the fields that we used to have of flowers, just condominiums all over now. There's no room for the flowers anymore."

This painful interview epitomizes important land-use history for Californians, noting that a modern person saw one part of the change. In 1944 Nicolus Hanson also mourned the changes he had seen over six decades in the Sacramento Valley:

> This and future generations will never see this great valley as I have seen it, blanketed most beautiful with thousands of acres of the most beautiful wildflowers. . . . And oh! The beautiful bouquets of wildflowers we gathered when we were a child. The fragrance never to be forgotten. . . . All these beautiful pictures of nature have been erased from this Valley for ever by civilization.

While reading this description during a workshop, I noticed that one workshop participant was also weeping. Lost as I was in the images evoked, I was startled for a moment by her tears and remembered that I too had them once. They probably dried while I engaged in what I now think of as a modern version of the flower dance. It has two parts, first, the fact that I can still "go flowering," and find much to experience, and second, the fact that I enjoy a sampler of the old splendor recreated on my own land.

Trips to the flower fields inspire the restoration gardener to plan new combinations, simple things perhaps not thought of before. Orange California poppy occurring randomly in a field of goldfields deepens the yellows with its richer color, or goldfields gleam against a background of white sand. And the combinations are infinite. In Bear Valley in Colusa County, I find groupings new to me: pink wild onion with light lavender Ithuriel's spear, dotted with pink *Calochortus venustus,* meadowfoam, and baby blue eyes. Or I find the opportunity to more closely observe a single species.

It is good to sit with owl's clover *(Castilleja exserta* subsp. *exserta),* each individual plant with three or four blooming stocks, their soft bristles set at different heights. Each flower head includes many shades of purple to magenta, darkening in the middle where they are shaded, lightening at the tips. Backlit, the white dots that make the owl's face aren't visible. With the sun at your back, the white dots appear. No species is simply purple or yellow, but sun, shadow, and placement deepen and lighten color.

THE SECOND PART OF THE MODERN FLOWER DANCE

The horticultural choreography of soil preparation, seed sowing, and general watching over necessary to grow California annual wildflowers is no different from that required by any annuals that are sown directly in the ground. In earlier gardening days, the knowledge and practice of sowing annuals directly in the ground was more frequently a part of the gardener's repertoire; the advent of six-packs and greenhouse-grown "starts" has rendered gardeners less likely to understand the factors involved in broadcast seed sowing. Like all seeds, wildflowers need freedom from the competition of weeds and sufficient moisture to germinate and make early growth, with concomitant protection from aboveground pests, such as deer and rabbits, and from belowground pests, such as gophers. As in the wildlands, results will be different every year. For the kinds of masses that are reminiscent of our great flower fields, direct sowing is appropriate. Growing wildflowers in pots yourself, or purchasing container-grown starts, can give you the opportunity to extend the season by tucking them into borders or bare spots in the gardens through the seasons, and into decorative containers as well.

TRICKS OF THE TRADE

My tears had quietly dissipated while I was glorying in what remains, which still is considerable, and figuring out how to enjoy a blooming display close at hand, managed by my own hand. I will go far to ensure that this phenomenon has a chance every year to act itself out in my restoration garden, so that it doesn't end when I come home from the flower fields. Baby blue eyes give way to Chinese houses followed by clarkias, which bloom by my steps. If I choose, as I frequently do, sequential sowings (even as late as March), they give me in summer the combinations of spring. Sometimes the harbingers of summer—clarkias and linanthus—mingle with early spring species sown in early spring, and the novelty of these combinations is entrancing.

Our fostering of California quail and the increased numbers of songbirds that come to our garden have changed the way I plant wildflowers. One December I watched bemused as a mixed flock of songbirds was joined by a quail covey in the area of my wildflower sowing. Curious to see how thorough they would be, I left them free access, with the result that not a single wildflower appeared that spring. Wildflowers appeared elsewhere in nooks and crannies, corners and crevices of the garden, but not in the big open dinner plate that was meant to be my flower field. Impressed by the birds' seed-finding abilities and happy to share seeds by the millions in other parts of the garden, I still wanted my flower display. Ever since that spring, I cover the seeds with a reusable plastic material, floating row cover. Permeable to rain and sunlight, this product, developed for agricultural crops, can be reused year after year. Once the seedlings are four or five inches tall and no longer tempting, the floating row cover is removed, dried, and stored away for later use.

Loss of my dog-and-cat gopher-hunting team some years ago also shifted predation in the garden. Without the free flow of coyotes through the town, gophers, with their valuable soil aeration abilities, are out of balance. Though they play an important part in providing good seedbeds in the churned up soil they leave at entrance holes, gophers can also claim responsibility for a wildflower field that never materializes (as well as the demise of many native bunchgrasses). To make things difficult for them and give myself a weed-free palette for wildflower sowing, I employ a technique borrowed from permaculture called sheet mulching. I start a new flower field or flower trail by mowing weedy grasses and other undesirable species closely, then laying several thicknesses of flattened cardboard boxes over the weedy area. Compost and soil is layered over the cardboard to a depth of four to five inches. Wildflower seed can then be sown directly; the cardboard seems to eliminate gopher predation until it breaks down. It's a no-till, no-herbicide way to prepare the soil for seed sowing. Weed seeds are not brought up from the seed bank, soil texture is not challenged by rototilling, and rhizomatous weed species are not given an advantage through mechanical breakup and spread of their rhizomes and tillers. Many retail stores break down boxes for subsequent removal and are glad to have them removed by customers.

DESIGN USES OF ANNUAL WILDFLOWERS

Landscapers can sow wildflower seed in disturbed soil to give clients something to enjoy while slower-growing perennials or shrubs are taking hold. In one situation, my clients' house was perched on a steep coastal bluff. Told that the cliff was unstable, they resorted to an elaborate engineering maneuver to protect their home, which left them with bare soil at the top of the bluff. Still holding the lower part of the bluff was a tough assortment of coastal scrub and coastal chaparral plants, and we agreed that our best bet was to duplicate that plant palette at the top of the bluff, where such plants had previously grown.

After installing the plants, a lot of empty space was left, and here we sowed wildflowers as placeholders until the shrubs grew to their ultimate widths. Sowing annual wildflower seed around the young plants produced a stunning show the following spring and gave these apprehensive clients pleasure and hope. The use of low-growing species, such as baby blue eyes, tidy-tips, goldfields, five spot, and red maids, allows for the healthy growth of young shrubs and perennials. Leave a one-foot ring of bare soil or mulch around each young plant to ensure that the wildflowers do not crowd or shade out the slower-growing perennials and subshrubs.

Looking for wildflowers is never a total loss, even in a less than munificent wildflower year. Much else always remains to be enjoyed, yet there is an unavoidable jolt associated with wildflower deprivation. How doubly welcome then the flowers of my home ground, whether tucked away in nooks and crannies or in effulgent masses. Returning reluctantly from my wildflower adventures, I am consoled by my own always interesting, always surprising backyard testimonials to this California phenomenon.

How lucky we are to live in this beautiful state and to have an opportunity, garden by garden, to deepen our sense of home through a studied, careful, playful, and spontaneous inclusion of California's wildflowers. Though the context is unknown to me, I concur with the sentiments of Wintu dreamer Mary Kenyon, whose song was sung by Edo Thomas for recording by the anthropologist Cora Du Bois in 1929.

From the old camping place
Comes a flash of flowers.
I love flowers.
Give me flowers.
Flowers flutter
As the wind raises them above.
I love flowers.
Give me flowers.

I divide my friends into those who have seen the glory and those who as yet have no idea of it, no inkling of what it is to wade right in, be surrounded and taken over, the faint trail of your footsteps quickly erased. Those who have seen it are changed forever. A poem on the subject by Wilma Elizabeth McDaniel, an eighty-six-year-old Okie poet who writes of the Central Valley and its people, says it well.

The Flower Lover

The visitors had been talking
about baseball, but somehow
Uncle Bart switched off to the
subject of flowers
jumped right in with, "Folks
if you really want to see a sight

take that Yokahl Valley Drive
why the roadsides and low spots
are live with monkey flowers
little yellow monkey faces
And beds of buttercups
And if you take that drive to
Woodlake
the lupine will knock you out
shimmery shiny blue alive
And poppies just popping gold.
It's a calling card from the Almighty
I tell you from my heart."
And his son tried to shut him up
"Dad, not everyone is as crazy
as you over California wildflowers."

I speculate that those who are not as crazy over California wildflowers haven't seen the remaining good places. Remaining unmoved at Table Mountain, Bear Valley, the Carrizo or Arena Plains would be a challenging task. In another poem, "First Spring in California, 1936," McDaniels describes the Okie reaction to the first poppy bloom, saying, "The women wept in wonder." Jefferson Mayfield's mother wept for their beauty in 1873. Juanita Centeno, a resident of the valley of the flowers, wept for their disappearance. Many tears have been shed, one way or another, over wildflowers.

In order to achieve this experience, it helps to plan ahead. Keep track of the rainfall and clear your calendar for April. Call the Wildflower Hotlines. (A list of phone numbers is given in the notes.) Let nothing interfere. Right now, it's one week into April, and I feel a melancholy descending upon me because I haven't yet seen their pretty faces in profusion. I know my health will suffer if the appropriate chemical responses aren't elicited. It's true that not everyone is as crazy for wildflowers as the poem's Uncle Bart, but it was pretty crowded in Death Valley last year (over the phone, the ranger said, "Don't come"). Joshua Tree National Park too was well attended.

And I am now a regular performer in our modern version of the flower dance, a horticultural rite enacted annually for several important reasons. To honor this threatened phenomenon, to flex my horticultural muscle, to nourish and continue the pollinator relationships, and to draw people in to our ecology and our flora—for all these reasons, I take part in the dance. And also for a moment that occurs reliably every spring.

On a day of sun between rain, when the air still holds moisture but the sun is hot, a subtle fragrance rises from the steaming earth. Momentary ascendance over the weeds has been achieved, and my battle with them, astonishingly enough, has become a dim memory, a forgetting that is a testimony to the power of the moment of which I speak. Resting my head on my crossed arms, from the corner of my eye I see my wildflower planting, the whole cast of characters, each with its own pollinator, type and taste of seeds, special fragrance, color, shape, relationship to the earth, its moment in the sun. And then, it's all worth it. It's all very much worth it then.

17 The Weed Dance in Modern Times

53% of California's most invasive plants have horticultural origins.

Sarah Connick and Mike Gerel, "Don't Sell a Pest," 2005

THE "YOU'LL BE SORRY" PLANT

A friend left Woolworth Garden Center carrying *Ipomoea acuminata,* variously known as "Blue Dawn Flower" and "Indian Morning Glory," in a one-gallon container. He wanted something colorful for his front yard, and the rich blue flowers of this non-native perennial morning glory looked like just the thing. The appealing young plant, which the nursery label described as "spreading well," already had numerous large blossoms and lush green leaves.

As he walked toward his car, he noticed a woman coming his way, heading toward the nursery. Her eyes lit upon the plant in his arms, and she veered slightly in his direction.

As they came abreast, she leaned toward him and said softly, "You'll be sorry."

Two years later, he remembers that woman and her cryptic comment, as *Ipomoea acuminata,* now dubbed the "You'll Be Sorry" plant, has invaded and smothered the junipers in his front yard, jumped the fence, climbed power poles, and, finally, made itself at home in his neighbor's yard. Nothing in the neighborhood is safe from it, not the Sunset Magazine Gardens, the Zen Gardens, the English gardens, the food gardens, or the native plant gardens. Creeping along as a ground cover for awhile, till it finds something, tree, shrub, fence, or telephone pole, to climb, it out-competes even the redoubtable English ivy, another better-known invasive plant still widely available in the nursery trade. Though it sometimes seems to be gone, it never is. Its rhizomes may be hiding under the juniper or behind the raspberry bushes, or lurking in the ground, biding their time.

Should nurseries be allowed to sell this and other overly aggressive plants? The California Invasive Plant Council, CAL-IPC, thinks not and is working with an organization called Sustainable Conservation to promote the voluntary cooperation of nurseries in avoiding these species. Though nurseries may resist being constrained in what they sell, the public can have an influence in this arena, as was demonstrated recently when a spate of letters to Home Depot resulted in the elimination of pampas grass from its nurseries.

THE WEED-FREE ZONE: A NEW DEFINITION OF A GOOD NEIGHBOR

Learning that 50 percent of invasive species in wildlands began as horticultural introductions, and after years of watching individual plant choices impinge on whole neighborhoods, I began to wonder: Should your neighbor be allowed to plant a species that encroaches on your yard? Vegetation issues between neighbors have not received much public attention; perhaps a new definition of being a good neighbor might include the avoidance of plants with poor manners.

In my town we have a volunteer-run organization called the Community Mediation Board. It serves as a sounding board for issues between neighbors, the premise being that it is preferable to bring problems out in the open in the presence of neutral, trained facilitators rather than to burn in silence. I am one of their best customers; invasive plants are the issue that brings me to the table.

I was dismayed when I realized that capeweed *(Arctotheca calendula),* a perennial listed as one of seventy-eight invasive species of greatest biological concern by the California Native Plant Society, was growing in my neighbor's yard and heading my way. Frequently planted as a ground cover in median strips, capeweed, with its sturdy rhizomes moving steadily through the soil, has characteristics that could easily smother my beloved and hard-won herb layer. Though not known to set fertile seed, its runners were already climbing my neighbor's vegetable garden and destroying her lawn. Still, not knowing how it got there, she considered it a welcome guest, low-maintenance and long-blooming.

After a number of years of watching capeweed move toward my garden, and after conversations with my neighbor about the problem proved unproductive, I called the Community Mediation Board and arranged for a meeting. I was curious about bringing vegetation issues to this venue. Guessing that my neighbor might remain unwilling to remove those pretty yellow flowers, which proved to be the case, I arrived prepared with a compromise. I asked for a "weed-free zone."

Acknowledging her right to grow in her garden whatever she liked, including a plant listed as one of the seventy-five most invasive species in California, I requested a six-foot-wide weed-free strip between my garden and hers. Since capeweed can send out runners six to thirteen feet long, a zone thirteen feet wide would have been ideal, but I settled for six. Hoping for a new definition of a neighbor's responsibility, we began work on the practical description of this border between our properties. The area was cleared of vegetation, landscape cloth was stapled to the ground, and wood chips were placed over it. It made for a tidy little border; it was a beginning.

MAINTAINING THE ZONE

Suppose your neighbor does agree to a weed-free zone between your yards. Accordingly, she removes cape ivy, capeweed, vinca, morning glory, English ivy, passionflower vine, or any of a number of other vining species, from her side of the fence, from whence the pest plant formerly spread to your yard. You regard her compliance with pleasure, till you notice that before it took place, the plant in question was able to take hold on your side of the fence, where it is now growing. Soon it will spread back to the other side of the fence. This important moment is your opportunity to set the standard for speedy removal, and to do so before it sends tendrils or rhizomes from your side to hers. Demonstrating compliance quickly boosts your credibility and dedication.

The vine is now gone from both sides, but it still lurks in no-man's-land, under the fence. From there, it may twine up the middle of a double fence and make a stand exactly in the middle. Somebody must take responsibility for this borderland, extricating each little tendril from between the wires or the slats or anywhere there is a space between the cement and the wooden posts. To forestall these problems, place cardboard, black plastic, landscape cloth, old carpet, or other impermeable surface under the fence, to prevent resprouting in this hard-to-reach area. Cover with wood chips or other mulch to weigh the material down and keep the area sightly.

This technique of sheet mulching, borrowed from permaculture, works well for many weedy species, particularly those with bulbs and rhizomes that can easily break off to form new, even more vigorous plants. Driving from San Francisco to the Oregon border along Highway 101 is a weedy experience, featuring that disheartening trio: fennel, teasel, and Harding grass. They occupy the zone between the asphalt and the fenced pasture, field, or development with a zombie-like consistency. No niceties of field characteristics apply to them;

they grow everywhere and will move with alacrity into all kinds of situations beyond the freeway fence. After years of sheet mulching these and many other species in my and many other gardens, with considerable success and relatively little labor, I now find myself mentally mowing, then sheet mulching, all the way to Crescent City. Would it be so difficult? At the very least there is enough recycled cardboard from all the big box stores between all the little towns to get a start.

Maintenance of the weed-free zone must take place on a continuing basis, as long as the species in question continues to grow in your neighbor's yard. If you are lucky, it may be that after years of maintaining the zone, your neighbor will agree that it would be easier to eradicate the species in question entirely. Public education regarding the harm that invasive plants do is beginning to have an effect. The property may be sold, and a whole new opportunity for neighborly agreements arise, in which your previous weedy experiences make you the knowledgeable old-timer. In any case, good weed-free zones make good, or at least, better, neighbors. Rural counties sometimes have laws penalizing landowners who allow agricultural weeds to grow on their land. Urban, suburban, and exurban land needs that kind of attention as well.

THE BIOLOGY OF WEEDS

During your discussions, your neighbor may reasonably request a definition of what makes a weed. For a long time, in gardening circles, a weed was defined as "a plant out of place." There was thought to be nothing inherently weedy about a plant, only that human preference and desire so classified it. Weed biology, now that billions of dollars are spent controlling invasive species in the United States, gives us a different picture.

There are many ways to be weedy. Some are obvious; others are less well understood. A weed may have plant parts that smother other vegetation, as is the case with cape ivy, kudzu, or passionflower vine. It may have long-lived seeds that are dispersed by mechanical or animal means, as with Scotch broom and pampas grass. Or underground rhizomes may give it an edge, as with capeweed. And some weeds have multiple strategies for their spread.

The botanist John Klironomos of Guelph University in Canada tackles the question from the point of view of what makes a plant "not weedy," that is, what it is that holds in check the population of a given species. Research on the relationships between pathogenic soil microorganisms and plant roots indicates that soil organisms affect the ability of plants to gain territory. Plants growing in places where they originally evolved may be kept in check by a mix of mycorrhizal components that promote growth and pathogenic components that suppress it. Klironomos demonstrates that some relatively rare plants grow more vigorously in soil that is not from their native clime than in soil from their home turf where the soil is replete with pathogens that coevolved with the given plant species. In other words, some plants that leave home have the potential to become weedy because no soil pathogens keep them in check. Freed of those constraints, they can become invasive in new territory. Other invasive plants do equally well in both home and new soil.

This complicated story translates for the gardener to a visceral knowledge of how weeds work, as we untangle vines, dig roots, pull annual grasses, and remove poison hemlock and fennel plants. William R. Jordan III says,

> It is action, the language of nature, that enables a practitioner such as a farmer, a hunter, a scientist—or a restorationist, whose work incorporates the experience of all three—to come closer to nature than a person can ever come through pure observation, or through words alone.

My friend who inadvertently introduced the "You'll Be Sorry" plant has convinced neighbors on all sides to cooperate in the total eradication of *Ipomaea acuminata.* He has sweetened the deal by agreeing to replace the old fence that is the staging ground. Though we deplore their presence, weeds teach us the complexity of the system, engage us completely, and require us to interact with our neighbors, including them in the weed dance in modern times.

NOT IN MY MINUTE

My daughter once asked me how our garden will do when I am gone to my reward. What she wants to hear is that after all these years, after this gargantuan effort, it will take care of itself, because it is "natural."

I wish I could reassure her on this score. But African veldt-grass (*Ehrharta),* which wasn't here when I arrived, is waiting under the porch to take over the places where our local clarkia grows, or where California phacelia is reseeding. My neighbor's capeweed is heading this way, but cape ivy will probably win that contest. Seedlings from the Chiapas oak brought here by the retired botanist down the street are appearing with alarming frequency, crowding out coffeeberry, slithering up through coyote bush. My daughter's interests lie elsewhere; it may be that she won't be inclined toward the devotion necessary to hold these borders. Though I have my hopes, I can't know how things will shake out, or which perspective will prevail in this town.

Once I attended a conference where indigenous Californians discussed efforts to keep their languages alive. In some cases, the tribes had only a handful of fluent speakers, even as few as one or two. One woman taught herself Rumsien Ohlone from old recordings, speaking it with nobody but herself. But the language won't die in her lifetime. As another speaker said, "Not in my minute, it won't."

Which is all any of us have, of course: our minute. I know that for now, coyote bush continues to be a generous companion to yerba buena, mugwort, and woodland strawberry, to bushtits, quail, and tachina flies. These things, and the life they manifest, won't disappear from right here, right now.

Anyway, not in my minute, they won't.

18 Do You Talk to Plants?

> Plants play powerful roles in human life—as food, narcotics, for mentors, healers, sedatives, tools, shelters, mood-makers. Our evolutionary continuity with them is profound: we share bacterial ancestors. But our perception of them as presences is limited by their immobility and form, a patience bordering on indifference. They are true beings whose otherness is so profound that it tunes and tempers our instincts for cover and comfort and protected observation. Their "intentions" appear to be at once more general and more subtle than those of animals.
>
> Paul Shepard, *The Others: How Animals Made Us Human,* 1996

Spring is also the time of vibrant reactions to plants; some thoughts on the topic of interspecies communications seem appropriate. In the course of my work with native plants, I am not infrequently asked "Do you talk to plants?" A satisfactory answer has not been easy to find. On the one hand, if I say "No," then I feel somewhat drearily limited in the nature of my plant-human interactions. Perhaps it will be inferred that I do not consider plants worth talking to, or inversely, that they do not so consider me. Or that I am too deaf to hear them, or too dull-witted, stodgy, or mechanistic in my thinking. On the other hand, if I say "Yes," I may be dismissed as anthropocentric, a wishful thinker, or, worse, a one-sided conversationalist.

I wondered how others in my field who have spent their lives interacting in one way or another with California native plants would answer this question. I thought I would ask in spring, since that season is well known for the intensity of plant presence. Everyone laughed at the question, and every single person wanted to answer it, in spite of its New Age implications, its fuzzyheaded desire to transcend our human limitations, to give other species and ours equal status, a hopeful suggestion of cross-species intimacy. Several replied but did not want to be quoted. Most were open-minded: Arvind Kumar, an active member of the Santa Clara Valley Chapter of CNPS and organizer of the South Bay Native Plant Garden Tour, said, "I don't talk to plants, but I respect those who do."

David Fross, the owner of Native Sons Nursery in Arroyo Grande and co-author of *California Native Plants for the Garden,* replied, "Do you mean out loud?" Then he went on to say, "Let me put it this way. I keep it to a whisper." Dave described his encounters with voices heard while exploring the Klamath region of California and Oregon. "The plants speak to me more than the geology and the rivers because that's how I was introduced to landscape, through the plants. I'm always thinking in the framework of plants and their edaphic influences." When asked if he spoke to plants in his nursery as well as to plants in the landscape, he replied, "Not as much when they are in little black pots and I don't know what their fate will be."

M. Kat Anderson, an ethnobotanist with meticulous research skills, replied, "No, I sing to them." Lillian Vallee, a restorationist, author, and teacher, said, "I sure do. Absolutely. Is there anyone who knows and loves plants who doesn't talk to them? That would be weird, I think. You talk, touch, mutter to, and chide them." When she engages in exotics removal she keeps up a running commentary, which she insists is a dialog. While removing poison hemlock from the San Luis Reservoir with her students from Modesto Junior College, she queries: "What the hell do you want here? What is it that you want? Can you go away and let other plants have a chance?" When asked if she minded being quoted, she said "My students already know I'm crazy."

Craig Dremann of Redwood City Seed Company talks to entire ecosystems, and twice in his life he has received detailed answers. When asked this question, he responded, "Please note that my pamphlet *The Redwood Forest and the Native Grasses and Their Stories* is written entirely in quotes. I took dictation directly from the plants."

An indigenous friend describes it this way: "Plants put thoughts in your head." They put thought of their needs in our head, through visual cues, the way the cat fixes me with an unblinking stare until I have determined what it is—water, food, or a change of locale—that it requires.

Conversation between plants and people takes on a more practical application when the question arises, "How did people learn which plants were edible and medicinal and which toxic?" Random experiments, frequently ending in mortality, are difficult to imagine, though in *Indians of the Oaks,* Milicent Lee recorded a story about an old woman who finds a mysterious large round fruit and is willing to make the sacrifice of eating it first. She breaks it open, samples its pink flesh, and, prepared to die, is the first of her tribe to eat a chunk of sweet watermelon.

Some people who write and talk about plants assert that if you are tuned in to plants, they will tell you what they are "good" for. They describe ways of learning the virtues of plants from the plants themselves. Tom Brown, wilderness survival teacher, describes his tender approach to learning about plants. Sitting next to them, prepared to spend time with them, he takes delight in their attributes. Inhaling their fragrance, he notes the complexities of their flower and leaf structure. By way of noting all their particularities, he sends respect and friendliness their way.

Joseph Bruchac, an Abenaki Indian from New York State, describes a common theme among Native American stories, a time when the people were harvesting animals in an irresponsible and disrespectful manner. In the Iroquois and Cherokee versions of the story, the animals retaliated by giving people diseases, which rampaged out of control. Things grew so difficult for the humans that the plants felt sorry for them, and from their bodies and their spirits they made medicine, for every disease a cure.

Patrick Orozco, a Rumsien Ohlone in California, tells a story he learned from his grandmother about another kind of plant-animal interaction from which humans learned. Once a family ate acorn that was improperly prepared. The agony they experienced from the tannins remaining in the meal was so painful to their father that he left the house and went down to the creek, where he saw a bear. Returning to the house, he retrieved some of the improperly prepared acorn meal and brought it down to the creek. The bear ate it and became as miserable as the people.

The man continued to watch. Eventually the bear looked around until it found yerba buena *(Satureja douglasii),* whose roots it dug and ate. Relief was immediate. Plant lesson learned from watching the bear, who knew plant medicine—this is a frequently encountered indigenous observation about bears.

Upon reflection, I have found a way to express something of what I feel regarding talk with plants. I put myself "in the sphere" of plants. I consider their colors, forms, textures, fragrances, their life histories, the places where they choose and do not choose to grow. And the names given them by people who name plants, both the unknown "folk" and the trained taxonomist. The potential for reproduction in the garden, as well as in the wild. The surprise of seeing a specific plant here and not there, or there and not here. What it likes or requires or can tolerate in the way of soil, exposure, and moisture. The sense of gratitude for its survival, whether in my garden or in the wild. A speculation as to its possible pollinators, or the insects and other herbivores who sample its leaves and seeds. These quick thoughts, as well as my personal history with *that* plant, conversations I may have had with friends regarding *it,* all swirl around the encounter, as I take in its presence and attend to its aspects. Something ineffable passes between us, as I suspect it does with all gardeners, which the herbalist Stephen Buhner calls a "sharing of soul essences."

For many years I experienced a powerful urgency around plants encountered in the wild, a compelling need to proclaim their beauty, the miracle of their presence in this place or that, and to enable their propagation by gathering their seeds. The most frequent cry I heard from the plants was, "See me." Did the insistent pressure that has accompanied my work with plants come from the plants themselves or from information, acquired elsewhere, regarding the many threats that lurk about their continued existence? Was this urgency that has driven me throughout my adult life the product of a wordless exchange that could be called a silent conversation?

A diviner trained by the Dagara tribe, a small, famously spiritual group residing in Burkina-Faso in West Africa, was the most recent person to ask me this question. "Do you talk to plants?"

"Not in English," I replied.

"Plants don't speak English," she said.

19 Rock Knows

Features for the Restoration Garden

The Garden Conservancy has once again enlisted nearly 450 gardeners in 26 states to share their gardens and, in turn, help support the Conservancy's work to preserve America's exceptional gardens. Lush plantings, tranquil resting spaces, and stunning vistas are just some of what you'll find when you visit these spectacular gardens.

Web site, The Garden Conservancy, 2002

The marvelous collection of gardens on the free, self-guided Bringing Back the Natives Garden Tour range from Jenny and Scott Fleming's fifty-year-old collector's garden to several that are newly installed, from five-acre lots to small front gardens in the flats, from locally appropriate to the horticulturally available suite of natives from throughout California, from Walnut Creek and Livermore to the Bayshore, and from gardens designed and installed by owners to those designed and installed by professionals.

Web site, Bringing Back the Natives, 2005

Some spring mornings make you feel you have awakened ten years younger than when you went to bed. You must hurry outside to see what's going on in the garden. On one such day I roamed my garden of coastal California native plants, enjoying baby blue eyes, noting pink flowering currant, and standing in silence in front of Douglas iris. Suddenly I remembered, with a jolt, that in one week my garden was to be part of the Garden Conservancy's Heritage Garden Tour. One hundred and fifty people would be making their way to our town to view five local gardens. Four are either estate gardens or gardens where many resources have been expended on hardscaping, and plants from all over the world are enlisted in the aid of the gardener's vision.

A kind of madness can possess a gardener when his or her garden is about to be part of a garden tour, and this morning I became momentarily deranged. I lost sight of all the modest and splendid ways of this garden that dance the line between restoration and gardening. Designed partially by jays and ants, maintained and pruned by quail and deer, its charms suddenly seemed so low-key as to be negligible in the realms of high garden art. In my embracing of the simplicity of this flattish one acre of sandy loam with no rocks, not trickling streams, and no views, I have avoided much in the way of recognizable garden features.

On this particular morning, I experienced a competitive horticultural surge. Partially stimulated by fear, it dimmed the glow of the day. I got out the tour prospectus and read the descriptions of the other gardens. Like a well-groomed poodle in front of show judges, each garden appeared replete with features: fountains, sculptures, trellises, terraces, rockwork, pools, ponds, waterfalls, gates, fences, and paths made out of unusual materials. My description was entirely plant-based. How could my garden impress those whose pupils had just been dilated by plant spectaculars and oddities from round the world amid the splendors of rock, brick, and tile in skillful array? In order to serve the cause and further the love of native plants close to home, I needed a garden feature, and I needed it fast. I knew what to do: get a big rock.

The day before at the lumberyard, a rock had caught my eye. It was a flattish rock, about three feet by four feet, in shades of cream and gray, slightly dome shaped, with pleasing angles and facets and a bit of moss. I liked this rock; it was a goodly rock. The presence of such a rock would denote painful effort, the danger of crushed fingers and toes, slaves to move it into place, trouble and expense. Set by the pond, it might give me an easily recognizable garden feature.

Till then I had resisted the addition of rocks to my garden because no rocks naturally occur here. There are rocks to the south of us and rocks to the north, but right here, on this marine terrace, no rocks. Dig in this sandy loam and you will find nary a pebble. Usually I delight in this buttery soil and have tried to make a garden that honors the marine terrace we have been given. Yet nothing sets plants off as well as rocks. What gardener doesn't find appealing the thought of tucking richly blooming ground-hugging plants into crevices and crannies in gray granite or shiny basalt? Deep blue against weathered gray. Rich yellow against creamy sandstone.

We gardeners long for that which we do not have, and that which we do have must be supplemented by more, or removed altogether, but it is seldom sufficient in and of itself. The excitement of last spring's flowers must be increased by more spring flowers next year. Spring flowers must be succeeded by summer flowers, and fall flowers, and also winter flowers. If we have oaks, we'd rather have redwoods; if we have redwoods, we'd rather have meadows. Flat places should be given berms, while hillsides are terraced to have flat places. Where there is no water, we will put water. Where there is water, we will install drainage. Where there is sun, we want shade, and where shade, openings to let in the light. Where there is no rock, rock shall be imported, and where there is rock, it shall be cleared away.

At the lumberyard downtown, landscape-quality rocks are available, brought here from rock farms in the foothills of the Sierra and rock farms in Sonoma County, distributed throughout this town to make walls, terraces, and garden features. Future geologists may find these additions confusing. Robert M. Thorson, in his geo-archaeological exploration of New England's stone walls, deplores their deconstruction and re-use as garden features elsewhere. Made from rocks found on site, they are an important historical and geologic record of the particular rocks farmers found close at hand at a certain point in time, at a certain place. To disrupt this record to provide a garden feature elsewhere is a loss of valuable information. For similar reasons, I have resisted the lure of California's rock farms.

But the garden tour is an event that can undermine the most deeply held convictions. I feared that my garden would look featureless to those garden tour attendees whose eyes have become attuned to glamorous and compelling garden features, which won't help the cause, the promotion of native plant horticulture. Garden-related businesses rejoice when they hear news of upcoming garden tours, knowing that the gardeners included on the tour will be knocking at their doors.

NAMING GARDEN FEATURES

But is my garden really so bereft of features? As I strolled around, I thought perhaps my garden description in the tour prospectus lacked imagination. Perhaps my garden has a number of features I have taken for granted, features I forgot to list. In order to be mentioned, features need names, so I spent the morning naming.

A shack hidden deep in the coyote bush, paint peeling and Dutch door partially hanging off its hinges, offers me a place for meditation and retreat. Hermitages being a classic feature of English gardens of a certain era, this authentically decrepit structure might be considered a charming place to encounter a hermit. Hermits employed specifically for the purpose appeared to visitors of English gardens in all their asocial splendor. A number of suitable candidates come to mind, but even without the hermit, the Coyote Bush Hermit Hut remains a legitimate feature.

We have no hot tub or outdoor spa, but the outdoor shower provides an exhilarating experience on a foggy day, when warm droplets from the shower-head mingle with cold droplets from our own low-lying clouds. On a sunny day, it's even better. Standing well within the embrace of a large red elderberry to receive the steamy blast, one is joined by birds alighting in its upper branches. This outdoor shower is frequently resplendent with spiderwebs, glistening with drops from the shower. Singing to the elderberry is optional. Some guests rave about our Elderberry Spa.

Also, we have the Bolinas Hedge, an attractive and effective windbreak and privacy screen composed of local shrubs. Across the path is the Quail Safehouse, a circle made of stock wire within which baby quail can be safe from cats. We have Deer Blood Oak, where a rutting buck left traces of blood on the cambium layer of the tree he rubbed his antlers on. Also, we have the Pond Scum Production Area, the pond that produces a lovely light green layer of algae, which can be dried for use as mulch, sponges, diapers, and other products. We have the Quail Dust Bath Spa, where the bodily imprint of a whole covey taking turns in the same quail-body-shaped hole has been preserved in the soft sandy soil. And nearby, we find the luxurious and verdant Quail Village, where soft cluckings are heard on a rainy day. Beyond that is Four Dog Oak, where canine pets of the past fertilize our largest oak, which at only twelve years old displays an eerie early maturity, the result of their nutrient input.

Now that they were named, features uniquely appropriate to the restoration garden emerged in their splendor, and I saw them for what they are, markers that honor the interactions we foster here.

(NATIVE) GARDEN OPEN TODAY

Had the Heritage Garden event been one of the magnificent native plant garden tours that have started up in recent years, I would have been at ease and might not have felt the need for a rock. First conceived through conversations on the Internet, the idea for a very different garden tour took shape in May 2003, with 1,200 people signing up to receive maps to a wide range of native plant gardens in the South Bay, from the small urban backyard, to the elaborate foothill estate. The first Going Native Garden Tour was a success.

Inspired by the South Bay experience, Kathy Kramer of Alameda County and the East Bay Urban Creeks Council took a slightly different approach to organizing an East Bay native plant garden tour for May 2004. An experienced event organizer and fundraiser, Kramer received funding to attack the logistical problems on a full-time basis. In 2005, three years after the inception of the idea, the San Francisco Bay Area added a third garden tour in San Francisco itself, so that altogether the three native garden tours included almost 100 gardens and hosted almost 8,000 visitors for a total of 23,700 garden visits.

Native plant gardens in Los Angeles and Orange Counties can be viewed on tour as well. Santa Cruz County native plant gardeners plan a tour for 2006, employing advice and assistance from the now relatively experienced Bay Area groups. As Arvind Kumar, one of the main organizers of the South Bay tour, said to me, "Tell me this is not a movement."

Kumar believes that native plant horticulture is of critical importance to the achievement of the goals of the California Native Plant Society. He says that garden tours, lectures, and get-togethers that foster and encourage native plant gardeners are as important to the California Native Plant Society as planned hikes are to the Sierra Club. Gardening, the country's most popular hobby, is a way to draw the public in, an accessible approach through which to teach love of native plants and concern for the ecosystems of which they are part.

Garden tours have other advantages as well: they provide a shot in the arm for horticultural businesses, a boost that native plant businesses can frequently use even more than the average nursery. Attendees are generally courteous and nondamaging, leaving hosts tired but exhilarated. Gardeners then discover the wonders of the day after, when the garden, brought to a rare state of perfection for its public appearance, can be enjoyed solely by the gardener and family.

THE ROCK ARRIVES

Still, thoughts of the rock kept returning till finally I gave in, made a phone call, and ordered it, arranging delivery for the very next morning, only six days before the big event. I then became extremely nervous at the thought of the arrival of the 1,200-pound object. To reassure myself, I visited the lumberyard to meet with the rock again and found that I still liked it. I sat on it, examined it from all sides, and made a rough sketch of its approximate shape. On one side, a blunt point projects out; I figured this could hang over the pond, providing a perch for thirsty birds. I tried to think geologically, which is difficult since, as I said, there are no surface rocks in this area. Back home, I walked around the pond, viewing from all angles where the rock could go and how it could be oriented. I considered plantings around it, light and shadow, possible animal use.

One rockwork expert, Phillip Johnson of Berkeley, recommends that the horizontal lines of each rock be placed parallel to the horizontal lines of the other rocks, an idea that makes good geologic and aesthetic sense, but I have no other rocks. Finally, I made a drawing of the best way it should go and returned to the pond to dig out an eight-inch-deep hole roughly shaped like the rock.

GOOD IN A DIFFERENT WAY

The next day two stonemasons arrived with my rock. Though they seemed nonchalant about the whole process, I averted my eyes as they pulled their small cart, rock teetering on top, through sticky monkeyflower and dune tansy to the pond, where I awaited them, drawing in hand. After a cursory glance at my sketch, they began to tip the rock off the cart, aiming it to fall into the hole I had dug.

Suddenly things took a different turn, as the rock seemed to develop a mind of its own, lurching sideways and falling in a completely new direction, away from its prearranged position. Both stonemasons leaped out of the way and waited for the inevitable. There was nothing anyone could do, no way to influence the trajectory groundward. The rock made a primal "Whuuf" as it hit the ground, and fine dust rose into the air around us.

At first I was dismayed and disappointed, for the rock had ignored my carefully dug hole, my idea about the projecting bird perch, my sense of its relationship to bedrock. The stonemasons were silent, probably glad to be intact. The question hung in the air between us, "Is she going to make us move this rock again?"

I looked at the rock in its chosen position, and the thought crossed my mind that it looked pretty good. Walking around it, I saw that, while not in the position I had in mind, this location created some possibilities I hadn't thought of. And so I gave up on my plan. I made adjustments, which in land management circles is called "adaptive management."

Next morning the rock looked as if it had always been there. Exciting new planting opportunities, unforeseen by me, presented themselves. (Of course, none could possibly be realized in time for the tour in five days. Indeed, nobody on the tour even gave the rock a second glance.) I planned which rocky outcropping in the nearby wildlands I would model the planting after, which native ferns (bird's foot fern) would be carefully placed in shaded crevices, which deep blue wildflowers (California bluebells) would set off its pretty tones of gray and beige. Things didn't go the way I planned; they went another way, also good, in a different way.

DECISION MAKING

With so many factors to consider, decision making can be difficult for gardeners. Sometimes we cannot proceed for fear of making a bad or wrong decision, something that will have to be undone. The idea that there is one right way, one correct choice, is an idea that furthers the job prospects and incomes of experts. Experience is invaluable, but all decisions, whether made by experts or beginners, will have consequences both desirable and undesirable, both foreseen and unforeseen.

In my early days, not sure of the direction this garden would take, I planted coyote bush everywhere. I'm glad I didn't wait to figure out the perfect place for each plant, when in the meantime coyote bush has supported so much life. It has held the ground and smothered out invasive weeds, improved the soil with its root activities and its leaf drop, and been a generous host to insect life and bird life.

Now that the land is held in this way, I can begin to carve out the spaces and the rooms and the nooks with a different eye, an eye that it took twenty-five years to develop. Planning a new restoration garden, we begin with new eyes. We can't begin with twenty years' experience; that take on this particular piece of property lies in the future, to develop only after two decades of looking, thinking, and evaluating. We can cherish the hours, days, and years spent on our hands and knees, with trowel and spade, putting back what used to be there, or our best idea of what used to be there, hoping that, given a backbone of local natives, we can't go too far wrong.

When asked the best time to prune coyote bush, or whether or not to prune salvia, I remember again that certain strategies or timing or activities foster certain species. Sometimes I try one set of strategies and some years another. I have developed several ways of answering vaguely, with "weasel words" that include *a bit, some, probably, maybe,* to allow for the unknown, the complex, and the site-specific.

We may worry needlessly, thinking there is only one best strategy for planting wildflowers, one perfect placement of the rock. The natural world includes multiple strategies for survival, many ways to wait out a six-month-long dry period, to survive a flood, to get seed to germinate, or to prepare a wildflower bed. The rock teaches me again that there is not just one right way for things to work. There is no one "best." This, rock knows.

20 Where's the Clover?

The Real California Cuisine

In former times we gathered clover.
Carolina Welmas (Cupeño), *Muluwetam,* 1973

I'd like to give some of those tree mushrooms to you to try them.
It wouldn't kill you. Well, perhaps it would a white person.
Villiana Hyde (Luiseño), *Yumáyk Yumáyk: Long Ago,* 1994

California cooks have a strong interest in continually increasing the range of vegetables and meats available to us. With their innovative use of fresh and unusual ingredients, the chefs of California have combined the cuisines of many cultures to create the elastic, healthful cooking style called California cuisine. These cooks are on the lookout for new dishes, ranging ever further afield for novel herbs and vegetables, ancient grains, and unusual techniques.

Though novelty may be the motivation, current research into our eating habits and health history indicates that this emphasis on variety may be a good idea. The rise of cancer has been linked to the decrease in the diversity of food crops that, according to Stephen Buhner, began around 1900.

> In people, increases in cancer exactly parallel the decrease of diverse plants as foods and medicines. In 1900, for instance, more than a hundred different types of apples, fifty different types of vegetables, and thirty different types of meat were commonly found in markets depending on the season. . . . Many of these wild-gathered plants contained the multiple types of cytotoxic, antimutagenic, and cell-division inhibitory compounds regularly ingested in human diets prior to 1900.

Looking east, west, north, and south, our chefs and food mavens consistently ignore possibilities for incorporating the food that nourished the indigenous peoples of California and attaching to our Western diet the healthful greens and grains of the place we inhabit. To redress this problem, we might include foods that kept the native peoples well fed and healthy for millennia. Some, like Indian lettuce *(Claytonia perfoliata),* are easy to embrace. Other foods, such as the native clovers, are more challenging.

One summer we had a booth at the Big Time celebration at Kule Loklo, the recreated Coast Miwok village at the Point Reyes National Seashore. The day before, we roasted seeds of blue wild rye *(Elymus glaucus),* red maids *(Calandrinia ciliata),* and chia *(Salvia columbariae)* to offer to the public. Also, in the way we had been shown by Milton "Bun" Lucas (Kashaya Pomo), we roasted, ground, and formed into balls the fruit of California bay laurel *(Umbellularia californica)* to make *behe chune.* A Pomo lady came by to look at our plate of the small pungent seed balls and asked the telling question, "Where's the clover?" I remembered Bun telling us that bay nut balls should be wrapped in leaves of native clover, but I had never seen, or grown, indigenous *Trifolium* species. So I began to look for native clover, and they were not easy to find.

At the time, it seemed to me that native clovers held a place similar to the one native grasses used to occupy fifteen years ago: they're not frequently seen, considering how prevalent they once were; they've only recently become offered in the trade; and few talk about them. Their ecology is complex, they have been devastated by invasive species, and, like the seed of native bunchgrasses, their foliage and sometimes flowers were an important food source for the native peoples. And their reintroduction, or attempts at such, teaches much about the indigenous management of the landscape.

SHE FORGOT TO LOOK OUT FOR BEAR

Clover is frequently mentioned in early historical accounts, indigenous myth, and ethnographies. Dances celebrating the first clover harvest were widespread, and according to M. Kat Anderson, clovers "were a mainstay in the diet of many groups. Of the thirty-one native clover species in California, at least fifteen were prized as food."

In indigenous stories, daughters frequently announce they are going to gather clover, mothers protest, daughters go anyhow. Mothers never want their daughters to go look for clover. Its deliciousness creates the potential for intemperate behavior, for the fatal single-mindedness that causes the daughter to "forget to look out for bear." When one daughter does not return, her mother goes to the clover patch and discovers the bad news. Her daughter has been completely eaten by a bear, except for one drop of blood. Fortunately, that drop has tremendous potential, can create basketry designs, can even become a grandchild.

Native clovers, while nutritious and full of protein, also contain toxins. How then to account for the importance of its harvest? Joseph Chartkoff and Kerry Chartkoff, in *The Archaeology of California,* describe two kinds of land-use practices.

> When pioneers move into a new area, they select features or resources to exploit on the basis of their existing knowledge, not on the basis of the potential of the new environment. This can be seen in modern times in the failure of European-American settlers to make use of the plentiful food resources of California that Indians used so successfully. Millions of pounds of protein in the form of hard seeds and acorns, suitable for human or animal food at no cost to grow, go unused every year in spite of the fact that the knowledge of the resources exists and that Californians pay a great deal to buy human and animal food. Their adaptation remains significantly pioneer because circumstances have not yet required the newcomers to adapt themselves more closely to their new home. In the same sense the Paleo-Indians [of California] remained pioneer settlers in their new home until the end of the Ice Age, when environmental change forced them to adopt new approaches.

And adopt new approaches they did, facilitating the growth of hundreds of different plants. Fire, coppicing, and pruning were employed to help plants flourish. Equally important, Californian Indians figured out how to take advantage of the nutrients available in certain plants without suffering the ill effects of their toxic elements.

QUESTIONS OF TOXICITY

Every California schoolchild is exposed to the processing and eating of acorns, but acorns are only one of a number of plants with toxic principles that were turned into nutritious parts of the California Indians' diet. Fruit of the California buckeye, soaproot bulbs, hollyleaf cherry kernels, angelica root, shoots of cow parsnip, lupine leaves—all were either harvested carefully at a particular time or processed so as to maximize palatability and healthful nutrients. The ability to ingest datura (*Datura meteloides),* once used in initiation ceremonies, in diagnosis of illness, and to foretell the future, without suffering the mortality that still awaits those who, with less knowledge, try its use today, was a significant technological achievement whose discovery is difficult to imagine.

Ethnobotanist Sandra S. Strike states that "California natives used numerous plants to produce a laxative or purgative effect. This may have been because their diet included foods which caused gastrointestinal problems, or may have been indicative of a ceremonial or ritual requirement that required purging the body." Perha ps native peoples were able to combine different foods in such a way as to render edible that which would otherwise be toxic or difficult to digest. For example, clover's degree of toxicity depends on species, time of year, plant part, and quantity consumed. Even as recently as 1902, in ethnobotanist V. K. Chesnut's day, one woman died from eating clover, so the toxic principles of clovers needed to be thoroughly understood.

Feasting on clovers in the spring was practiced by many tribes, including the Luiseño, Patwin, Maidu, Pomo, Wintu, Choynumni Yokuts, and the Sierra Miwok (reputed to have eaten at least eight different species). Of some species, only the preflowering foliage was eaten, of others, the flowers and seeds. Clover was eaten fresh, steamed, dried, mixed with salt, dipped in salt water, and combined with other foods, such as bay nuts. It was frequently noted that Indians relished the appearance of wild clovers in the spring, though some Europeans who observed Indians eating clover described this joyous event as "grazing like animals."

One example of combining foods to counteract or balance the deleterious effects of ingredients in the meal is the Wappo practice of eating bay nuts with green clover to prevent bloat. Having eaten roasted bay nuts at many open houses with no bloating problems, I conclude that the bay nuts are intended to counteract the bloating effects of clover. Other tribes recommended that the clover be dipped in salt water, which may also help bloating.

It's easy to imagine that much of this invaluable knowledge was lost, first of all, because it wasn't valued by Europeans, and second, because in many cases it was female knowledge, even less valued. William Least Heat-Moon proposes a third scenario:

> This passing of useful botanical knowledge from red people to white is the exception not just because whites often scorned such knowledge, but because medicine men and Indian women, the keepers of much of this knowledge, had little commerce with settlers, and also, surely, lore must have been deliberately withheld from a people taking away the land, so that the thieves got the big machine but not the operating instructions.

Though once abundant in grasslands north of San Francisco Bay, the native clover *Trifolium amoenum* was presumed to be extinct by 1984, according to the California Native Plant Society. Its common name, showy Indian clover, reveals two important characteristics. First of all, it is showy, which translates as "of interest to the gardener." Second, it is associated with California Indians, which might indicate that it was one of the clovers that were eaten, though Miwok Maria Copa Frias, when interviewed by Isabel Kelly in 1932, did not think it was the most delicious. (Some of the species identifications in these interviews, however, are suspect, and in this case, in another field note, the specimen identified as *Trifolium amoenum* is referred to as "better" than other clovers.) Much is lost in the mists of time. Kelly's interviews with Miwoks Tom Smith and Maria Copa Frias indicate that not all members of the genus *Trifolium* were eaten; her informants apparently recommended *T. fucatum* and *T. wildenovii.*

In 1993 a single plant of *T. amoenum,* about to be bulldozed next to a road near Occidental in Sonoma County, was rescued by Peter G. Connors of the University of California Bodega Marine Laboratory. Sharing his ninety-two collected seeds with another researcher at UC Davis (practicing restoration guideline no. 4—when growing special plants in the garden, hedge your bets—see chapter 26), Connors and his colleague grew eighteen plants.

The seed that was obtained from these plants made possible experiments by graduate students and other growers, who encountered the same problems that gardeners must reckon with. In our garden, I planted rancheria clover (*T. albopurpureum*) one afternoon, and by the morning of the next day, the plants were decimated. Native clovers are candy for snails and slugs. We now put clover grow-outs in our stock production area, with protection against snails, deer, and rodents, and keep some growing in containers, where conditions are easier to control.

Can we grow enough that we might experiment with eating clover? What tricks and techniques might be helpful to gardeners wanting to restore this important genus, so rarely seen now in our grasslands? The clovers, two-thirds of which are annuals, respond readily to simple germination techniques. The main consideration besides weed control seems to be snails. Growing them in bare dirt with no mulch or with a rock mulch, unpleasant to snails, may be helpful. Snail baits, from dishes of beer to virulent poisons, are readily available. Frequent nighttime snail raids could also be helpful. A new product that is non-toxic and can be scattered in the soil nearby remains effective till rain falls. Copper strips surrounding the growing area are repellent to snails and slugs. A combination of these techniques could allow the beautiful flowers and healthful greens to flourish again, so that some day modern Californian daughters may once more be tempted by the rare taste and helpful nutrients of native clovers.

Summer

21 Saying Farewell to Spring
Growing Site-Specific Clarkias

> The genus as a whole has been given the somewhat poetical name of farewell-to-spring, but the present showy species [*Clarkia amoena*] is distinguished by the specific appellation of herald-of-summer. In this latitude it distinctly deserves that name, for I invariably find the first seasonal blooms within a day or two of the summer solstice.
>
> Leslie L. Haskin, *Wild Flowers of the Pacific Coast,* 1934

When in the throes of spring, it's hard to imagine being ready to let go of its blazing intensity, so I am always surprised how enjoyable it is to relax into the summer season. The young quail partnering up and moving near cue me in to the change. In their adolescent boldness and innocence, they come close at this time of year, using the garden gates, fences, chairs, tables, and sculptures as perches, our grassy areas and wildflower fields as feeding meadows. Soon they will be making casual nests and laying eggs in California fescue and Pacific reed-grass clumps.

As I walk my garden in June, July, and August, I revel in the remaining wildflower bloom. Much is still flowering. Summer can be a floriferous time in the native plant garden, thanks to the herbaceous annuals and perennials, the forbs, our treasured "small things." In particular I enjoy the genus *Clarkia,* named after Captain William Clark of Lewis and Clark fame. It is notable within the family Onograceaea for a collection of characteristics that makes its members of great interest to the gardener. Many of the sixty-one annual species are easy to grow and adaptable to a variety of garden conditions. Since many parts of California have their own clarkias, growing local species is a way to have our gardens sing their unique, regional song. With remarkable reseeding abilities, they bloom at a time of year when the dominant theme in California's hills is golden grasses, with which they can coexist to a surprising extent. And they bloom long, eliciting praise from the gardener, who frequently enjoys them without any seed sowing, watering, or weeding.

Ringing countless changes on the color pink, clarkias come in pale pinks, deep pinks, and pinks that are almost red, saturated lavenders, pale lavenders, and lavenders that are almost pink. The striking dabs of color that account for a good part of the picturesqueness of this genus, the so-called "blotches," "wedges," or "zones," are found in different locations in different species, either at the base of the petals, partway up, or at the tip. They can be wedge-shaped or irregular, carmine, baby pink, and also pure white, set against the whole aforementioned range of pinks. According to Lester Rowntree, an entirely white clarkia from the California-Oregon border, *C. puchella* var. *alba,* was once available.

Some clarkias are diminutive, like Davy clarkia *(C. davyi),* hiding in the grass, while others are tall, like mountain garland *(C. unguiculata),* rising boldly above it. Many of these taller species are prolific seeders and can compete with many of the non-native grasses that occupy so much of their territory (not including Harding grass). Some grow in full sun while others prefer a bit of shade, and yet others can handle both conditions.

One such versatile species, and the clarkia perhaps most frequently seen in gardens, is farewell-to-spring *(Clarkia amoena).* This species had a long history of human association even before the famous Scottish botanist David Douglas sent seed back to England where, under the name "godetia" (though several other species are also known as "godetia"), it became common in English gardens. Before that, *C. amoena,* as well as a number of other *Clarkia* species, was an important seed food for indigenous peoples. Elders of many California tribes remember that fires were set to enhance the production of wildflowers with edible seeds, prominent among them being farewell-to-spring. The seeds were gathered, cleaned, roasted, and ground. According to M. Kat Anderson, they were sometimes made into soup or, in the case of the southern Maidu, into "large loaves of mashed, cooked seeds." Among the Sierra Miwok, the man's name *La'uyu* is said to imply, "Mashed farewell-to-spring seeds adhering to lips when eating," an indication of how important this seed food was.

In England, California's farewell-to-spring was selected, hybridized, played around with, and returned to us as showy as any petunia. *C. amoena,* the horticultural version, can now be found in the cut-flower trade, sold in nurseries in six-packs, and in some years covering gaudily resplendent acres at Stanford University in Palo Alto.

Other members of the genus *Clarkia* in close association with human use include *Clarkia breweri*i, granted "uncommon" status (meaning not frequently encountered) in the *Jepson Manual.* The fragrance of this small flower is so remarkable that its chemistry has been disentangled and incorporated into the scent-making processes of the world's perfume industry. Few other clarkias are fragrant. Another species, mountain garland, is in my experience one of the world's longest-lasting cut flowers. The relatively small pink and purple flowers of mountain garland are spaced along an upright stalk, which lends itself to dramatic presentation in a tall, narrow vase. Grown in large containers (we use wooden boxes four feet square), where it receives regular water, it can bloom for an inordinately long time as well, even, along the coast, in October.

THE CLARKIAS OF CALIFORNIA

My local clarkias, of which I grow about six species, never fail to surprise me with their pleasing and varied floral displays at a time of year when I am still, though I should know better by now, simply not expecting the gorgeousness they present. One year, I was able to add Raiche's red ribbons *(Clarkia concinna* subsp. *racheii),* a rare annual with deeply lobed petals, so delicate and exquisite that it merited star status in a pot in my sunroom entryway. Many more possibilities, within the relatively local, tantalize me. Gardeners in Sonoma County might try growing Vine Hill clarkia *(Clarkia imbricata),* which, though endangered, is in cultivation. Featuring beautiful fan-shaped petals, "lavender shading paler near middle, with large wedge-shaped purplish red spot above," this federally listed wildflower makes a stunning bedding plant. The Sonoma County Milo Baker Chapter of the California Native Plant Society has successfully extended the population of Vine Hill clarkia through grow-outs.

In Mariposa County, a gardener might grow the Merced clarkia *(C. lingulata).* In the San Jose area, fairy fans (*C. brewerii)* can be grown in a lean potting mix, then transplanted into the garden. In San Francisco, it would be a coup to have in a prominent place in the garden a small ceramic container abloom with the delicate Presidio clarkia (*C. franciscana),* another endangered clarkia whose population has received enough care and protection that seed might someday be available at local California Native Plant Society plant sales and wildflower shows.

THE HILLS WERE MADE BRIGHT

I have a special fondness for one of our local clarkias, *Clarkia rubicunda.* I like the way this clarkia explodes into view upon first sight of it growing in profusion on a certain rocky roadside just before the Steep Ravine turnoff on Coast Highway 1. On a sunny June or July day, I may be compelled to park at the turnoff and walk back up the road to spend time with these clarkias. They are accompanied by a rich flora of other granite-growers in retreat from the weedier flat areas, but on this day the clarkias are the main act.

Lustrous in the sun, the rich pink of the cup-like flower is set off by the dull granite boulders whose fissures it inhabits. The deep ruby-colored collar at the very base of the pink petals is a design that could grace wallpapers, draperies, and silk scarves.

One year, after enjoying this granitic display, I returned home to find that a seedling of *Clarkia rubicunda* was growing out of the gray cement foundation of my house. The flowers blooming against my gray house reminded me of my original inspiration on the Steep Ravine granite. When holes in sidewalks and other concreted areas bloom with wildflowers rather than the usual weeds, backyard restoration gardeners can rejoice in the sense that the land under their care has become so thoroughly saturated with native plant seeds that even the cracks and crevices are turning up native species.

Where I live, we are at the northern limit of the range of *Clarkia rubicunda,* and also at the southern limit of the natural range of *Clarkia amoena.* Part of knowing about my place is recognizing that I live in this "clarkia transition zone." To the casual eye, the most obvious difference between the two species is the location and size of the "blotch." In *C. rubicunda,* this intensely colored mark settles deeply at the base of the flower, where it's referred to in the *Jepson Manual* as a "zone." In *C. amoena,* a red or dark purplish red patch occurs near the middle of the petal.

I became interested in providing seed of the "real" *Clarkia amoena,* the one still found on rocky hillsides in northern California. Generally of a softer pale pink than the commercially available form, with shocking pink blotches halfway up the petal, this wild form is delectable in masses on hillsides. It became my goal to collect seed from populations of both these species on private land. I first asked permission from the landowner and made sure that each population was large and vigorous, covering at least 250 square feet, and occurring in at least three other locations nearby, so that a small seed collection would do no harm. I picked one flower stalk of each species, slightly before maturity, allowing the narrow inch-long capsules to fully ripen in a stainless steel bowl or paper bag. When the capsules begin to dry and become woody, they split open along a seam from the top. There is no explosive dehiscing as with poppies or lupines. In fact, extricating the seed in coastal areas where summer fog retains moisture in the capsule can be a process requiring pounding with a rolling pin. Or you might try a technique used by one California Indian tribe: pull up the entire plant, dowse it thoroughly in a stream, then leave it to dry on a sunny rock, after which the capsules open readily.

In October we sowed the seeds in flats, then transplanted the small seedlings of both species to 4-inch plastic pots in January. We placed these pots in the garden in two ways, both in large wooden boxes and directly in the ground. Clarkias began to bloom in June and, as of this writing, are still in bloom at the end of August. Though not as opportunistic as punch-bowl godetia *(C. bottae),* which turns up everywhere, *C. rubicunda* seems willing to stick around, given a little help. Recently, I noticed it growing across the road in a vacant lot, where perhaps long ago it once flourished.

WHAT'S IN A NAME?

For the "cup half-empty" people, *Clarkia amoena* is called "farewell-to-spring" in the *Jepson Manual.* For the "cup half-full" people, it is elsewhere called "herald-of-summer" and also "summer's darling." Other *Clarkia* species, such as *C. gracilis* (described in Sharsmith) and *C. williamsonii* and *C. purpurea* (both in Anderson), are also known colloquially and locally as "farewell-to-spring," as the genus provides many regions in California with summer bloom.

The need for botanic names is well demonstrated by the genus *Clarkia,* where one text (Kozloff and Beidleman) calls *C. rubicunda* "godetia," while another *(Jepson)* gives that name to *C. amoena* subsp. *amoena. C. rhomboidea* is called "tongue clarkia" in Kozloff and Beidleman but "common clarkia" elsewhere *(Washington Wildflowers). Jepson* names *C. exilis* "the slender clarkia"; in Kozloff and Beidleman, *C. gracilis* subsp. *sonomensis* bears this vernacular name. And the *Clarkia* species included under the rubric farewell-to-spring, are many. How many encounters are required to make a name "common"?

I frequently encounter species given common names I have never heard spoken or seen written elsewhere. Yet somebody somewhere must have spoken to somebody else about this plant, at least once. Leslie L. Haskin describes this phenomenon in the introduction to his book *Wildflowers of the Pacific Coast,* written in 1934. Haskin calls himself "the first to write in popular form of the flora of this region."

> Many of them (the wildflowers) also, have no popular names. . . . Many of our most interesting plants have, indeed, no Anglo-American history, for our flowers have not yet fully woven themselves into the lives of the people. We must look to an older nation for their true story, and for folk-tales and traditions we must trace out those vague stories that, even yet, though fast passing, are to be found, here and there in the lodges of the Indians.

Although no "common" name, defined by the *Jepson Manual* as "a colloquial name in common usage," is listed for *Clarkia rubicunda* in the *Jepson Manual,* the CalFlora database gives it the appealing name "ruby chalice clarkia," which is in the vernacular. Many places in coastal California must have once been made bright by both *C. amoena* and *C. rubicunda,* places now shrouded in weedy grasses, fennel, teasel, and the whole suite of weeds. Though I've never heard *Clarkia amoena* referred to as "summer's darling," the name reveals that somebody, somewhere, appreciated this plant. Were it commonly known by such an affectionate name, how could we let it disappear? It would be equally hard to let go of "ruby chalice clarkia" and "winecup clarkia" *(C. purpurea* subsp. *quadrivulnera).*

Though inconsistent and unreliable, common names help plants become a familiar and comfortable part of our lives. Common names confer value, indicating that the plants are the object of human attention and esteem. With the intimacy that comes through horticultural interaction, and the validity conferred by usage, common names of clarkias from our local floras may help our gardens become places of enriched conversations. As centers of surviving species, they may be refugia from which, some day, some part of California may be returned to the splendor of summer's darling and the brightness of ruby chalice clarkia.

22 The Preservation of Small Things

It remains a fact that the average American township has lost a score of plants and animals through indifference for every one it has lost through necessity.
Aldo Leopold, "The Farmer as a Conservationist," 1939

While engaging in the play of gardening, the gardener can also be a student and savior of the most threatened part of the California flora, our native forbs. I think of these plants, both annual and perennial, as "the small things," the non-woody herbaceous species under two feet tall. Many of them, like the clarkias of the previous chapter, are *of concern* to me by virtue of their growing habits: their closeness to the ground renders them vulnerable to a sea of invasive plants. Many of them, far more in number than are currently available in the trade, are critical habitat for insect pollinators and other fauna and have the potential to be fascinating garden subjects. Once the shrubs and trees are established, many native plant gardeners eventually find themselves drawn into this perhaps more advanced arena of native plant gardening: the preservation of small things.

Sometimes these plants surprise us with the conventional garden roles they are happy to assume: as specimens in hanging baskets and ceramic containers, such as Indian pink *(Silene californica)* and yerba buena *(Satureja douglasii),* as filler between bricks, such as cudweed (*Gnaphalium* spp.), as green and flowering ground cover in arid areas, such as hayfield tarweed (*Hemizonia congesta* subsp. *luzulifolia*) and the coastal form of the California poppy (*Eschscholzia californica* var. *maritima*), as dried flowers, such as evergreen cudweed *(Gnaphalium californicum),* and as food, such as yampah *(Perideridia kelloggii),* to name a few, a very few, of the lesser-known scores of possibilities.

THREATS TO THE HERB LAYER

It's possible to drive the back roads and freeways of many parts of California and enjoy the impression that the plant communities of California are relatively intact. Trees and shrubs, both next to the road and in the distance, appear reassuringly abundant. They are rarely old-growth, or even second- or third-growth, but the species palette is still in evidence; oaks, bays, buckeyes, willows, maples, conifers, and dozens of other woody species are all still present and accounted for. Less apparent is the destruction of California's herb layer.

For a variety of causes, many native forbs that are not listed under the state or federal endangered species acts still face local extinctions. Their vulnerability depends on a complex of characteristics, including height, adaptability to a variety of sites, the need for specific mycorrhizal fungi, the way in which they are pollinated, and the form of seed dispersal.

Most pollen-producing plants have pollen that is too heavy for dispersal on the wind, so they depend on animals, most frequently insects, but also birds and bats, to move pollen from one flower to another. The more threatened the animal pollinator, the more threatened the plant. Pollinators that migrate, such as hummingbirds, bats, and certain butterflies, are more easily disrupted by development, pesticides, and freeways. The host plants must be available in contiguous patches or pollinators may move on before they flower.

Here are three particular stories that demonstrate the scenario of which I speak. None of the three species involved is protected by federal or state regulations.

Eighteen years ago, I collected chalk buckwheat (*Eriogonum latifolium)* from a population surrounding the Bolinas lagoon. Chalk buckwheat, with its felted silver leaves, drumstick flowers in pale pink or white, and tidy, low-growing habit, hugs steep, rocky places. It is an important larval host plant for a number of butterflies as well as a significant component of our native rock gardens. But now the original population is no more; the area is a mass of pampas grass and teasel.

Twelve years ago I collected (with permission, on private land) a small amount of seed of peppermint candy flower *(Claytonia sibirica)* from a large population that seemed vigorous and here to stay. Today that lovely mass of peppermint candy flowers, its delicate white blossoms thinly striped with pink, is completely gone, replaced entirely by periwinkle (*Vinca major),* which was not even present when I made my collection.

Eight years ago I collected seed of ruby chalice clarkia *(Clarkia rubicunda)* from a nearby bluff, rocky enough that it seemed little else could grow there. Subsequently, fennel (*Foeniculum vulgare),* lurking nearby along the road and sending its seed into tiny cracks, moved in from the roadside and obliterated that shining stand.

These three species have refuge in my garden now, where they more than sing for their supper, in terms of beauty, ease of culture, wildlife value, and historical rightness. As I drive past the places where they used to be, it helps a bit to know they have a place outside my back door.

GRIST FOR THE GARDENER'S MILL

It's no chore and no penance to introduce these beauties into the garden, all of which have surprised me with their qualities. One example is peppermint candy flower, which is related to both our ubiquitous Indian lettuce, *Claytonia perfoliata,* and the treasured spring beauty *Claytonia virginiana* of the East Coast. Growing as far north as Alaska, candy flower is also known as Siberian spring beauty (and in Kodiak it is called "rain flower"). The leaves are added to salads, soups, and sandwiches, and the flowers are used as a garnish. Some of its close relatives have sturdy tubers that are considered a delicacy.

A perennial, it dies down in late summer and fall, then reappears and blooms from February into May or June, or even later on the coast. It fills that sought-after niche of a plant that grows in part sun or deep shade, requires no water, and flowers prolifically. Challenged with dense willow shade, it blooms happily and returns year after year. Its leaves, though not quite as mild-tasting as miner's lettuce, are edible when young. I grow it in containers also, where it blooms for months in the spring. Seed that falls into the container from the spring flowers will sprout early in the fall and sometimes bloom yet again before cold sets in. A blue ceramic container sets off the pink stripes of its modestly sized but pleasing blossoms.

This year candy flower spread into the area where I grow Point Reyes checkerbloom (*Sidalcea calycosa rhizomata)* and together, the two flowering perennials make a spring statement of pink prettiness. Though candy flower is not a listed species, Point Reyes checkerbloom is. An invaluable ground cover for somewhat shady areas as well as a vernal pool plant, it's an example of what makes California one of the world's twenty-five "biodiversity hotspots." The organization Conservation International lists California as one of the biologically richest places in the world, containing as it does such a high number of endemic species, plants that grow nowhere else. At the same time, with all its legendary beauty, California is classified as one of the four most degraded states in the United States of America. Through growing these species in the garden, we make a small step toward ensuring their survival.

THE QUESTION OF SEED COLLECTION

Seed collectors in the wildlands operating without either traditional or legal restrictions could conceivably be devastating to native plant populations, a most undesirable scenario. It is, of course, illegal to collect any part of any listed species, except in the path of a bulldozer and with permits. Yet these same species rarely have protection from fennel, teasel, Harding grass, and all the other rampant invasives that can be equally destructive. It's a conundrum; on the one hand, there would be no stock for native plant nurseries if collections had not been made at one time or another, on the other hand, threatened populations need all the seed they can get.

The opposite of love, it has been said, is not hate but indifference. At this particular juncture in time, the greatest danger to our native forbs may truly lie in indifference rather than in hate or greed.

There is a small but increasing number of success stories in which human intervention, through collection and grow-out, has preserved and even increased populations of rare species. Vine Hill manzanita (*Arctostaphylos densiflora)* and showy Indian clover (*Trifolium amoenum),* two Sonoma County species that almost became extinct, have been brought back from the brink by carefully monitored reintroductions.

For the gardener, there is plenty of opportunity to enhance local populations through use in the garden without touching on these challenging issues. Leaving aside all listed species and concentrating on the many species that, though not considered threatened, desperately need our attention, the gardener can add continuously to a rich, diverse plant palette right outside the door.

THE PRESERVATION OF REALLY SMALL THINGS

In my dream neighborhood, in the secret heart of some backyards, is a place where neighbors show off their horticultural skills by providing a home for the more obscure herbaceous species, those with flowers too small to find a place in most wildflower books, those whose virtues may have so far gone unremarked. These are the species that in another, less progressive, day might have been categorized as "not gardenworthy," their flowers dismissed as "insignificant." In my dream neighborhood, some of the gardening energy once devoted to producing the largest, gaudiest, and showiest flowers is now given to producing those with the most minute, delicate, and subtle blooms. Some, the smallest of the small, have amazing features that must be seen with magnification to be appreciated, so a hand lens is kept nearby in the garden. Jeffrey Caldwell, landscaper-ecologist, points out that many in this category are native annuals, bounteous seed producers with significant wildlife value. Being the smallest, they are the first to go.

Native plant nursery people, given encouragement, are frequently eager to grow these treasures of the connoisseur. They are happy to have a reason to experiment with the obscure and subtle and to find an appreciative customer response. Margaret Graham of Mostly Natives Nursery in Tomales wrote on the label describing coast plantain *(Plantago maritima),* "You can be the only one on your block to grow this plant."

In a world dominated by plantains from Europe, it is satisfying to foster our native plantains. This one is found on coastal bluffs growing with soft sage *(Artemisia pycnocephala)* and sea pink *(Armeria maritima).* To ensure its initial survival, we planted it closely in a wide, shallow terra-cotta container, where this interestingly succulent plantain came into its own, sending up numerous straight brown seed stalks. Everybody who saw it in bloom wanted it; alone in the garden, it attracted little human attention. Another native plantain, the annual *Plantago erecta,* is one of only two larval host plants for the bay checkerspot butterfly. Even more diminutive than the maritime plantain, its survival always seems a marvel, and it can be found both in the desert and in coastal prairies where weeds have not yet invaded. A hand lens reveals its intricately tiered flower stalk.

THE USE OF CONTAINERS FOR FORBS OF CONCERN

I frequently grow those forbs that are of concern to me in ornamental containers, both to make it easier to observe the plant's decorative qualities and to ensure myself of a source for seed and cuttings, a common practice in the trade. In a container, it is easier to monitor pests, weeds, and water needs, but a container is not just a replacement for growing these species in the ground. It allows me to single out each one, saying: This plant is special; this plant is worthy of our attention. It's a luxury available to Californians, whose moderate climate makes possible the long-term survival of plants out of the ground. Back East, unless the plants can be given protection in greenhouses, container growing ends with the arrival of frost.

In containers, surprising things sometimes happen. Annuals may defy their projected life span, like the *Clarkia amoena* that turned into a semiwoody shrub, living and blooming for two years in a terra-cotta pot. Or the containers of mountain garland and ruby chalice clarkia that have been reseeding for years. Though annuals, they became a perennial presence, blooming in the spring, going to seed, receiving water for the other things in the pot, and sometimes germinating and blooming again before winter.

When housed in an appropriate container, their qualities are held up to our view. I have frequently been surprised by characteristics that are displayed once a species is grown in a container and receives the irrigation that pot culture requires. A quick look through my garden reveals the following examples.

Coast lotus *(Lotus formosissimus)* from a nearby meadow of California oat-grass trails elegantly over the sides of the pot. In a shady spot, redwood sorrel *(Oxalis oregana)* in a wide, shallow container stays green and flowers for months, quickly filling the pot with its fresh, clover-like leaves. *Linanthus montanus* from Mendocino County has been reseeding for several years in its own container, demonstrating, as the members of this genus often do, how striking simple flowers can be. Pacific stonecrop (*Sedum spathulifolium)* comes in three colors: lawn grass green, silver gray, and purple; I keep all three around in pots, sometimes growing together.

Container plants will eventually need to be moved into larger containers or given fresh soil, though not as frequently as you might think. For humus-loving container-grown plants, I mulch with rich compost. For species that prefer lean soil, like *Lewisia rediviva* or *Clarkia breweri*, special mixes can be made or purchased. Container growing allows you to grow some species that don't appreciate your garden soil. Those vulnerable to snails can be more easily protected with copper strips or nearby snail bait.

Through container growing, seed and cutting stock is also protected until such time as sufficient seed has been produced to sow in flats or directly in the ground. Sometimes annuals like *Clarkia* or *Madia* begin seeding into the ground around them, or into other pots.

GROWING IN CONTAINERS

Following a few simple guidelines will ensure success with container plants. All except riparian or aquatic species must have good drainage holes in the bottom of the container to ensure the movement of water out of the pot once the roots are saturated. It used to be thought a good idea to partially cover the drainage hole with broken pieces of ceramic pottery; few bother with this anymore. It also used to be thought that ceramic pots fostered better plant growth than plastic, but the vast quantity of nursery-grown plants that have spent their lives thriving in plastic pots pretty much disproves that theory. Terra-cotta pots are for pleasure—for the feeling of solidity, antiquity, and tradition they bring to the growing of plants in containers.

Choose containers that will allow for a few years of unimpeded root growth, and indulge a bit with container selection. I allow myself a few new glazed Chinese pots and some standard terra-cotta pots every year. To begin, I empty old containers of soil, wash them, and place them on a table, with standard organic potting soil mixed with some humus from my garden nearby, and plants and seeds at the ready. The next part is delicious fun: deciding whether deep blue desert bluebells *(Phacelia campanularia)* should contrast with a rich mahogany-glazed pot or gently blend into a pot with a sky-blue glaze. Calculating how long it will take the silk tassel shrub *(Garrya elliptica)* to bloom in a container. Keeping an acid-lover like the native rhododendron in a giant Chinese urn where I can provide the soil it likes but would never receive in my garden.

Plants in containers benefit from shade at the end of the day, even if they are sun-lovers. The more crowded the root-ball, the greater the possibility of heat buildup that can kill the roots. Water in the morning by filling the pot from the top of the soil line to the rim of the container. Return five minutes later and do the same thing again. One of my old horticulture teachers used to delight in showing us how a pot whose soil seemed to be saturated with water after a single watering could be tapped out of the container to display an absolutely bone-dry root-ball. Mulching is also helpful to preserve moisture. Succulents and cacti and other desert species enjoy a stone or gravel mulch.

Having container plants to fill in at doorways, display seasonal bloom, and take up space where annuals have gone to seed allows much in the way of creativity, particularly in the small garden. Using them to grow our threatened small things expands the possibilities and adds deeper motivation.

A SAMPLER OF SMALL THINGS

My list is always changing, as some plants stick around and thrive, while others, for reasons of their own, disappear. Here is a sampler of other herbaceous species not commonly found in the trade, to which a backyard restoration gardener might give refuge in the garden. Your sampler of small things will be quite different from mine, depending on where you live and what interests you. Frequently, these are annuals or perennials whose seed I may hope to offer to the public and with which I am unfamiliar. While increasing stock and learning characteristics, I also provide my close-in living areas with enjoyable garden features. None of the species below is given threatened or rare status, either federally or by the state of California, yet their occurrence is infrequent enough to be worrisome, a sighting rare enough to be noteworthy.

Indian Pink *(Silene californica)*

This attractive perennial has tubular coral-colored flowers beloved by hummingbirds. In the wild, Indian pink exhibits what is called "summer dormancy," during which time the leaves die down to the crown. It often looks as though the plant has disappeared entirely. The most recent edition of the *Jepson Manual* includes for the first time horticultural information for some species. The appellation "DFCLT" (difficult) is attached to the description of *Silene californica,* possibly because of the complete aboveground disappearance of Indian pink in late summer.

Perennials that die down to the crown are in a vulnerable state while asleep. Weeds may move in right over them so that they awaken from their dormancy to find themselves smothered and overtaken. Overwatering can rot their sleeping roots, which may also be eaten by gophers. Too little water or too high heat can overtax their ability to resist desiccation.

In containers, where the plants receive extra care, dormancy can sometimes be overcome, and the plants may unexpectedly thrive without the rest we thought they needed. One specimen of Indian pink grown by a customer produced both a luxuriant spring and summer bloom. "Not difficult at all," she said.

Lester Rowntree's description of Indian pink is one of my favorites among her thumbnail plant sketches.

> Here on this hot slope you may find scarlet Indian pink. It is never wise to waste time in hunting this gadabout for it takes great satisfaction in growing in unexpected places and among strangely dissimilar associates. The thing is to try to forget it and then, suddenly, you will find it is there. It may nestle casually below pines or drop their long roots from sunbaked cuts, its root tips safely embedded; its flopping terminal stems are adorned with large flowers whose petals are unevenly split and divided. This *Silene californica* has some gorgeous color forms in shades of salmon and bright red, and I have sometimes found flowers of almost pure yellow.

Red Ribbons *(Clarkia concinna)*

The deeply lobed petals of this part-shade to full-sun *Clarkia* are entrancing in the woods or on road cuts, but it's a bit fussier than some of its close relatives and can disappear when conditions are not to its liking. A less munificent seeder than some, its seed is not widely available. In a container it can be coaxed and coddled, and the delicate shape of its deeply lobed petals seen close up and enjoyed. Rowntree touts its appearance "just as the greens are quieting and the tans and ecrus of summer coming on."

Checkerbloom (*Sidalcea* spp.)

The mallow family gives us many showy and hardy species, like *Lavatera assurgentiflora,* which, much appreciated by deer and gophers, survives many violent attacks to thicket vigorously, producing in abundance large pink flowers with large light green leaves. The related semi-woody *Malacothamnus* species will also thicket all over town if given the chance and are appreciated for their exquisitely shaded cup-like blossoms. All of the low-growing herbaceous *Sidalcea* are lovely, and most are perennial, but the pale pink *Sidalcea diploscypha,* with its delicate, ghostly appearance, is an annual. It's one annual that has not reappeared when planted in the ground here, so to further seed production and ensure its survival here, I also grow it in pale blue ceramic containers.

Goldback Fern *(Pentagramma triangularis)*

This diminutive fern is always a pleasure to find growing on mossy or rocky banks, for it thrives in a variety of situations, as long as the drainage is good. I keep it around in several containers, and it does not go dormant in this cooler coastal climate, though it will further south. One of my first experiences with Gerda Isenberg involved her showing me how to turn the fronds of goldback fern over to admire the spore pattern, the "gold" found in complex arrangement on the back. I treat this fern as a pet, cutting dead fronds back to the crown and repotting it as needed. In general, ferns can handle being fairly root-bound.

SUMMER IS THE TIME

Many herbaceous species bloom in the spring and summer and set seed before or during August, so the months of June and July are a busy time for seed collectors. This season is designated by the Central Pomo as the time of ripe seeds, or Umchachich-da. Gardeners in cooler climes may want to sow seed flats soon after collection, transplanting seedlings into four-inch pots by early fall and placing them in the garden during the rainy fall and winter. Those in the hotter summer parts of California will have greater success if they wait for cooler days before planting.

In either case, summer is an important time for seed collection of California's native forbs. It's a time when determination, focus, and a passion for biodiversity can converge into concerted attempts at the preservation of small things. While engaging in pleasurable gardening activities, the California backyard restoration gardener can play a satisfying role in the perpetuation of the herbaceous species of California's flora. In the *Jepson Manual*'s necessarily terse plant listings, the following horticultural comment is attached to the description of certain species not yet in cultivation that the author deems worthy of gardening attention:

"TRY."

23 The Pollinators of Small Things

In the *Handbook of Biological Data* of 1956, the Monarch is listed as a pest. It is (even more) difficult to discover the reason for this, since the larva of the Monarch eat nothing but milkweed, which in the same volume is also listed as a pest.

Jo Brewer, *Wings in the Meadow,* 1967

What the honeybees provide through sheer volume, the native bees provide in efficiency.

Mike Koslosky, "What's the Buzz? Native Bees," 2003

Our specialists don't lack for information about symbiotic relationships, but neither expert nor layperson holds dear what is known. The information we might read about this phenomenon exists outside of our lives and speaks of the world we no longer know from direct experience. Consequently, it has lost its power to inform and enrich us.

Joanne Lauck, *The Voice of the Infinite in the Small,* 1998

As far as I know, I have never had an insect "pest" in my garden. I glance at the many shelves of chemical sprays in the garden store and walk by quickly. To the backyard restoration gardener, a strange insect is usually just "a friend you haven't met yet," an important part of the ecosystem, to be admired and enjoyed. Liberation from the notion of insects as pests is not only a release from the time-consuming, expensive, and potentially destructive use of pesticides, it is also a propulsion into a fascinating piece of the world that even the most urban gardener can promote and enjoy.

Walk out your back door and there it is, in even the smallest urban garden—an ongoing party with partners chosen long ago. Sitting by flowers becomes a profound experience when you can watch these small creatures, here long before us, and know that you have furthered their mutualistic dance with native flowers. Wildflower viewing becomes an opportunity to observe the party, while fostering it makes the heart swell with the particular mix of emotions experienced by those engaged in some aspect of habitat gardening. No matter what the size of the garden, the preservation of small things brings with it the opportunity to encourage native insects. Researchers find that native plants are four times better at fostering native pollinators than non-native species. When I see bumblebees on my lupines or solitary bees on my meadowfoam, I know that I'm giving them the best possible food for their needs, the nectar and pollen with which they coevolved. Equally important, I know that the flowers they visit will be pollinated as efficiently and fully as possible, producing good seed crops.

Good habitat for native bees, butterflies, beetles, flies, and other pollinators is also good garden design. Planting the one, you create the opportunity for the other, thereby enhancing possibilities for the survival of both and providing immediate evidence of the positive role you play on this particular stage.

To foster bees, responsible for perhaps 85 percent of the world's pollination, avoid spottiness; the occurrence of many small patches of different kinds of flowers isn't good for the human eye or the bees. Patches of a single species should be at least sixteen feet in diameter to ensure that the effort of getting to the patch of wildflowers is worth it to the solitary bee. Bees forage most efficiently when they can remain in the behavior pattern associated with a certain kind of nectar and pollen extraction. Since pollen from one species will be wasted on the female flower parts of a different species, better pollination and more complete seed set will follow as well. No matter the size of the garden, strategies for creating habitat to attract native bees, butterflies, flies, and beetles are the same as strategies for backyard restoration.

Also recommended is the provision of a wide variety of species, at least eight different kinds, to provide sequence of bloom, as well as opportunities to attract different kinds of pollinators. A variety of flowering plants through the season, both those with simple, open blossoms, like asters and daisies, and those with more complex flowers, like lupines and salvias, provides for the tastes of a wide variety of pollinators, and people as well.

Planting numerous different bee-attractive species near one another will increase native bee use. Some hybrids currently in the trade, bred for size and showiness, are often low in nectar and pollen, and some may actually possess neither. They are flowers without the purpose that drove their evolutionary history, and a waste of a bee's time.

The social workings of the honeybee, and the delicious results in the form of honey, have so captured our imaginations and enriched our cuisines that many people are surprised to learn that honeybees are only one species of bee among many thousands. Books with titles including "the bee" in them (such as Maeterlinck's *The Life of the Bee* and modern novels like *The Kiss of the Bees* and *The Secret Life of Bees*) cement our ecological illiteracy by supporting the implied assumption that the European honeybee *(Apis mellifera)* is the only bee in town. Equally surprising to many is the information that the honeybee is not indigenous to this country and that, of the four thousand bees native to North America, about 75 percent are solitary bees, with no social life comparable to the complex and well-documented honeybee hive life. In California, our many species of solitary bees lead Californian lives that are at least as "secret" and equally fascinating.

The French entomologist Henri Fabre, in his day called "the insects' Homer," is one of the few who have sung the fantastical ways of solitary bees. In 1843, as a young teacher in the town of Carpentras, France, he took his students out to the flat, flinty, thyme-covered fields called *harmas* to teach surveying. Here he witnessed strange behavior from his students, who, ignoring their assigned tasks, walked across the flinty fields holding straws, then occasionally dipped down to insert them into the ground. Investigating, Fabre found that they were using these straws to break into the cells of a velvety black bee "who made clay nests on the pebbles in the *harmas*. These nests contained honey; and my surveyors used to open them and empty the cells with a straw. The honey, although rather strong-flavoured, was most acceptable. I acquired a taste for it myself and joined the nest-hunters."

Intrigued by this bee, Fabre acquired a text on insects, found therein listed the names of the naturalists of his day, "and, while I turned over the pages for the hundredth time, a voice within me seemed to whisper, 'You also shall be of their company!' "

That's what a solitary bee did for Henri Fabre, who then did much in return for the insects. He preserved his backyard *harmas* for their use, studying them and experimenting with them in unique ways that, while perhaps not entirely palatable to modern sensibilities, yet played a major role in bringing them and their ways to the attention of the general public.

Still, until quite recently the European honeybee has thoroughly eclipsed the critical role played by California's native bees in the pollination of California's agricultural crops. Though the native alkali bee *(Nomia melanderi)* has been well known to alfalfa seed growers of the Great Basin in Nevada and Surprise Valley in Modoc County, California, since the late 1940s—and orchardists are familiar with the activities of the orchard mason bee *(Osmia lignaria)*—the pollination activities of native bees didn't gain state and nationwide publicity until two species of foreign mites, the varroa mite and the tracheal mite, began to devastate honeybee hives. The much-admired social nature of the honeybee made possible the spread of these imported pests, while native solitary bees remain, so far, unscathed. Providing native bees with what they need so that they in turn can pollinate our commercial crops has become a matter of national importance, receiving significant media coverage throughout the country but particularly in California, which grows so much of our nation's food.

A modern-day entomologist, Edward S. Ross, is to me one of our current "insects' Homers," with the added advantage of being a skilled photographer whose pictures of insects attain unbelievable intimacy. At a lecture supposedly focusing on protecting our homes from termites, I found myself being introduced by Professor Ross to the world of the solitary bee. I learned that the bumblebees visiting my lupines may actually be bumblebee imitators, such as *Anthophora bomboides stanfordiana.* Like other solitary bees, wasps, flies, and even beetles, these imitators are harmless and seldom sting, banking on their resemblance to their fiercer cousins to protect them from predators. Look for *stanfordiana* (this species has no common name) in warm, sunny, well-drained banks, frequently by the sea. The presence of its individual nest chambers will be marked by "a turret of excavated mud." The adult dies with the onset of cold weather, leaving her offspring in their cells, which Ross describes rather cozily: "Throughout fall, winter, and early spring, all that remains of the once-bustling bee community are hundreds of fully fed prepupal larvae in their underground cells. The snug confines of their earthen burrows will keep the infant bees safe and dry throughout the winter.

BEES AND MY DREAM NEIGHBORHOOD

In my dream neighborhood, schoolchildren learn that the honeybee is actually an import from Europe and that there are many thousands of bee species native to this country that could benefit from our attention. They learn that in California we have more than 1,500 different native bees, the vast majority of which are solitary bees.

Their unique life cycle, though rarely if ever celebrated in story and song, stimulates the imagination in a Californian direction. Demonstrating faith in a future they won't see, the female bees devote considerable energy to carefully constructing little chambers in rocky or sandy cliffs, bare dirt, twigs, beetle tunnels, walls, or human-made nesting boxes. Using local, renewable resources—from leaves to wood to mud—to line the chamber, they are the original "green builders," working cleverly with their materials and safeguarding the health of the inhabitants. To these chambers, through numerous trips to flower fields, they bring pollen and nectar, mixing them with their mandibles into a homogeneous ball called a "pollen loaf," reputed by some humans to be not very tasty, unlike Fabre's bee honey of the *harmas.* When the mixture is just right, they lay their eggs on top, seal the chamber with more of the same cell-building materials, and depart, to die before their children are born.

Solitary bee eggs hatch into larvae that are immediately surrounded by evidence of their mother's thought for their future, the food mass upon which the eggs were deposited. The larvae begin to eat three or four days after they are laid and continue for a month. The sturdy structure their mother built for them should remain impervious to cold, heat, damp, mold, fungus, and a number of parasites, but in case it is not enough, each larva, after digesting and eliminating its food for a few days, begins to weave a very tough cocoon around itself. Within this protective shell, it metamorphoses into an adult solitary bee. A year after the egg is laid, it emerges, orphaned, but with many siblings.

BY THEIR WORKS YOU SHALL KNOW THEM

Some of the relationships between solitary bees and annual wildflowers, particularly species from vernal ponds, are *oligolectic,* that is, the bee requires a specific flower to survive, while the flower, though it may be pollinated by other "generalist" bees, receives the best-quality pollination services from that particular bee species. That bee's adult activities are limited to the bloom period of its pollen host plants, so perfectly are they matched. Vernal pool wildflowers, famously showy, are frequently pollinated best by specialist bees. My favorite example is meadowfoam, *Limnanthes douglasii,* one of the earliest wildflowers to bloom. Though a wildflower associated with vernal ponds, meadowfoam in the garden is one of the most adaptable, long-lived, and reliably reseeding of all the wildflowers. In masses it is surpassingly beautiful and has a faint, delicious fragrance. In a field with some natural moisture, it blooms first, along with

Indian lettuce, to be succeeded as the field dries up by lupines, baby blue eyes, tidy-tips, and goldfields. Ever since I saw it as the first flush of a local wildflower field, along with Indian lettuce, I have included it in our local wildflower mixes. Its presence clued me into the reality that this hayfield was actually once one massive vernal pond. It comes in yellow and white *(L. douglasii),* pure white *(L. alba* and *L. douglasii* subsp. *nivea),* pure yellow *(L. douglasii* subsp. *sulphurea),* and white with a faint rosy tinge (the rare Sebastopol *L. vinculans*).

My identification skills are not such that I can always be sure whether native bees have been in the neighborhood, but if I find good seed set on meadowfoam, which means that all four of the possible seeds are fully formed rather than two or three, I tentatively conclude that solitary bees have been around. Partial seed set indicates that non-native bees may have done the pollination work. Orchardists experiencing a cold spell that limits honeybee activity when fruit trees are in bloom can count on good pollination thanks to the orchard mason bees, who are more active in colder weather than the honeybee. By their works you shall know them. Though native pollinators' nesting requirements are complicated and in many cases not well understood, the pollination services they provide are significant, and likely to become more so in the future.

IN PRAISE OF BARE DIRT AND DEAD TWIGS

How can we help ensure the presence of native bees? First, by eliminating the spraying of toxic chemicals, second, by planting the native species with which they coevolved, in as large patches as possible, and third, by providing appropriate nesting habitat.

Almost two-thirds of our native bee species make their nurseries in poor, open ground. Barren cliffs by the coast or rocky road cuts inland are potential homesites, as are dirt paths, bare patches in the lawn, and constructed berms and mounds. Eschew the use of landscape cloth and leave some places free of bark mulch as well, to allow the solitary bee direct access to the earth.

Other native bees make nurseries in tunnels in dead wood, frequently ones excavated by beetles. A corner of the yard where the common garden practice of removing dead limbs is not so assiduously practiced, a place for the strategy of benign neglect, can be good nesting habitat for bees. Shrubs with soft, pithy centers are sought after by many solitary bees, so planting elderberry makes sense, and not just to bees.

The endangered valley long-horned elderberry beetle requires elderberry to survive; its larvae favor the soft interior of elderberry's twigs, through which they tunnel, leaving good runways that solitary bees can later partition off to make their individual nurseries. Indigenous flute and clapper stick makers also use hollowed-out elder branches for music making.

The sweetly fragrant white flower panicles of the blue elderberry, pollinated by bees and other insects, face the sun till the weight of the ripening berry crop causes them to turn over and droop toward the ground. White bloom on the blue black berries makes them appear pale blue, "elderberry blue." Many people prefer eating them when baked in pies or dried. Elderberries make small, intensely flavored raisins, loaded with polyphenols, those protective plant chemicals associated with plant pigments with which we evolved. The sun seems to sweeten the berries inland, but our coastal blues have a rich, unique flavor and make excellent pies. Many indigenous peoples across the continent ate of the genus's bounty. Dozens of birds love elderberries, and in some parts of California, mountain quail are expected to come down to the valleys and flatlands when the elderberries are ripe.

Ben Cunningham-Summerfield (Mountain Maidu/Turtle Mountain Ojibwa), who makes and plays the traditional elderberry flute, said, "Elderberry gives us the clapper sticks and flutes that help us give voice to our songs and to our stories." It seems that the muse, along with solitary bee larvae and valley long-horned beetle larvae, also resides in the elderberry.

24 Animal Assistants

Under the ceanothus and behind the shrub lupine is an almost inaccessible place of hard-packed bare dirt. With only four feet of headroom under the ceanothus, maneuvering in this area is difficult, yet it looks as though the ground has been carefully prepared for the sowing of wildflower seeds. How did this perfect seed bed get there?

QUAIL HELPERS

Every weedy plant that sprouted here has been diligently grazed by our garden helpers, the California quail. Their rigorous attention to this area is reminiscent of the livestock management strategy known as "mob-stocking" or "intensive short-term grazing." Large numbers of sheep or cattle fenced into a single pasture are moved out after two weeks to a month, not to return for six months to a year. During their short stay, they prepare the soil for new growth in the spring, churning the seeds into the soil, breaking it up, and recompacting it. Sometimes hay is strewn throughout the area where livestock action is desired. *Excitation,* or vigorous activity stimulated by fear of predators, is artificially produced to ensure trampling of seed into the soil.

In August I see nothing but bobbing quail bodies, no ground visible as the quail perform their own brand of intensive, short-term grazing. I have only to stroll among them to produce excitation. Then I know that in a few months, I will be able to sow wildflower seed with no other preparation. Of course, there are many other ways to do this, including manual weed removal, spraying of herbicides, burning, and rototilling, but in this area, the quail prepared the soil.

In his book *The California Quail,* A. Starker Leopold shows an aerial photo of the quail roosting area on the ranch owned by Ian McMillan. A bare ring around the area where the quail relished all the new green growth in the spring and ate the weed seeds in the soil is apparent. I am gratified to have such an abundance of quail that now I too can point with pride to my own quail-created bare ring. A question still remains: How can I combine quail foraging activity with producing the wildflowers, both annual and perennial, that I want to see here? One answer is floating row cover.

The quail are even more enthusiastic about native wildflower seed than they are about weed seeds, so after the sowing, I cover the area with floating row cover. Developed for row crops, this material lets in moisture and sunlight while preventing the quail from devouring the young wildflower seeds and sprouts. After experimentation, I have concluded that I must leave the floating row cover in place until the seedlings are four to six inches tall, at which time the quail lose interest in them. Removing it when the sprouts are only two inches tall is too soon. At that time, the little plants are delectable to the quail, a tasty tidbit.

Removing the harvest guard too soon or rather, too soon for my purposes but just right for the quail, I found the densely growing seedlings mown down flat to the ground. I was not in despair about this occurrence—or rather, I was in despair at first but quickly shifted to that kind of low-key acceptance of gardening events that resist my will that is such good practice for the rest of life.

A month or so later, it became apparent that the grazed area was recovering. As it turned out, the flowers bloomed their heads off for our midsummer workshop, several months later than they otherwise would have. I received this unexpected midsummer bonus as yet one more gift from *Callipepla californica.* I took the credit for this out-of-season floristic event, though in reality, concept, soil preparation, and sequence of bloom were provided by the quail.

DEER HELPERS

People frequently ask why the coyote bush looks so good in my garden, compared to the scrawnier representatives they see elsewhere. One of the answers may be found in the small pellet-shaped black scat that thickly litters the ground in some places. As well as enriching the soil, research shows that ingredients in deer saliva can stimulate growth.

Once plants are established, they may benefit from a certain amount of deer browsing. The key phrase is "a certain amount." Though deer can be benign or even helpful in the pruning of established plants, young plants will not enjoy their ministrations and may even be pulled out of the ground by tugging deer mouths. In all plant-animal relationships, timing and degree are of the essence.

Since there are now more deer in the country than were present before Europeans arrived, it is quite likely that wherever you live, deer populations are out of balance. Or they may not be out of balance today but they will be tomorrow. Deer populations, like quail, raccoons, ravens, and crows, all of which are well adapted to human habitation, rise and fall quickly.

Deer damage in gardens is a major concern in many parts of California. Through the years I have become less and less enthusiastic about recommending certain species as "deer-resistant." Show me a list of "deer-resistant species" and I will show you an almost worthless piece of paper. I worked on a garden once that was near a county park, prime habitat for an expanding population of black-tailed deer, and deer families could frequently be seen trotting down its suburban streets. So I was assiduously consulting the lists of deer-resistant plants. One such list put California fuchsia-flower *(Epilobium canum)* at the top of its choices of deer-resistant perennials. Accordingly, I included it in a planting in an area that received much deer depredation.

First and above all, these deer loved the *Epilobium canum.* They ate it mornings and they ate it nights. They ate it when we were present and when we were not. They ate it coated with various deer-repelling substances, and they ate it plain. Not believing the evidence of our senses, we replanted, this time giving each plant its own separate wire cage, a labor-intensive operation. The deer then pruned each plant to the silhouette of the square, carefully nibbling each shoot that dared to venture outside the perimeter of the cage. When we removed these cages, each plant was perfectly square. From this experience, I learned that *Epilobium canum* is a good candidate for the pruner's art, something I would never have guessed had not the deer shown the way. Good ideas come from the animals as they demonstrate new horticultural uses for the plants.

Deer populations vary in their vegetable preferences, and a given population will change what it considers acceptable food based on the season and on the menu. My deer may relish what your local population disdains. Even within a specific herd, tastes change throughout the season. A recent local barbecue included both pork from feral pigs rampaging in nearby wildlands and venison contributed by a neighbor who had harvested a deer in his garden. Upon dressing the venison, he found that the deer's stomach was full of rosebuds.

One morning I watched from my window as several deer moved through my garden. Later I went outside to retrace their path. I noted where they had eaten the tops of the coastal form of the California poppy, then moved on to nibble delicately at the tips of thimbleberry, then munched a bit of coast live oak's tough and spiny leaves, ending with the flower stalks of checkerbloom *(Sidalcea malvaeflora)* and alum root *(Heuchera pilosissima).* Only the checkerbloom was adversely affected by the deer's "little bit of this, little bit of that" ramble through the garden. After being trimmed, those alum root that had been browsed sent up new stalks and bloomed later in the season than those that were left unscathed, thus prolonging the bloom season and the opportunity for those pollinators that enjoy them. Like the activities of the quail, deer "pruning" serves to spread out bloom time so that the flowering season for a number of species starts later than it otherwise would. The unpruned bloom at their usual time, and that sought-after garden phenomenon, *sequence of bloom,* is provided free of charge by deer and quail.

I became used to seeing and musing about their delicate nibblings and the ways they combined foods. Perhaps it is the rare exotics, like tea roses, I thought, that elicit junk food–like responses, leading to binging. I felt good about my relationship with the deer, who were not a problem to me.

Things remained in balance for twenty years or so; I provided a bit of this and that for a varied basic diet, and the neighbors supplied the roses for dessert. Then another neighbor put a road and small house along a creek, a place previously hidden in willows where deer ate and slept unnoticed, in blissful privacy. Now the deer family is on the move, its sleeping and eating habits in flux. Overnight an equilibrium was lost, a loss that is a mere shadow in magnitude of the lost equilibriums of the last century and a half. Important grow-outs of clarkias, local nemophilas, coast lotus, and others disappeared overnight, and I must now fence that which has gone unfenced for thousands of years.

This change provides a hint, a mere taste of the disruption of previous centuries. I am reminded of the delicacy with which the Choynumne Indians supplied Jeff Mayfield's family with game, leaving venison and elk at their doorstep. These gifts came with an ulterior motive. In later years Jeff learned that the Indians had hoped to forestall the settlers' disruptive hunting techniques, which might send the animals far away and increase the difficulties of locating prey.

A WORD ABOUT DEER-PROOFING PLANTS

Numerous products are available to help gardeners protect their plants from hungry deer. Many techniques and implements are employed to deter and repel them, including substances applied to plants to repel by both odor or taste, such as lawn fertilizers, garlic, hot pepper, the urine of mountain lions, coyotes, and wolves. Scaring devices include motion sensors that squirt water, ultrasonic devices that produce high frequency noises, and dogs. Here's what works: the fence. Many choices are now available, including an unobtrusive black plastic fence. Traditional stock-wire fence is inexpensive and sturdy and has stood the test of time. It has the advantage of not using more of our forests, or of somebody else's forests. Hedges can quickly hide it. Note that deer will sometimes use shrubs to climb a fence, so keep a cleared zone outside your fence. Double fencing is also extremely effective. When deer are motivated to jump a standard six-foot fence, add another, parallel, fence four feet away. The second fence should be three feet tall and placed outside the six-foot fence. After jumping the shorter fence, the deer will not have the room for the necessary leap over the taller fence and will turn around and jump back out.

With fifteen to twenty-five million deer in this country, neither white-tailed nor black-tailed deer are in danger of extinction. We do them no favor to encourage them to expand their population, using our gardens as exotic California cuisine. In suburban and exurban areas, they lose their saving timidity, an attribute that may yet serve them well sometime in the future.

I watch a doe move past my window. An occasional car goes by, a dog barks. At each event, the doe, hemmed in between my house and the road, shudders and seems to consider desperate options. She is nothing but nerves, strung so tightly I wonder she can live with such sharply honed paranoia. Her stolen meal exacts a high price in anxiety.

ANTS

A cheery sight in mid-February on our native prairie and in containers as well are the flat yellow blossoms of sun cups *(Camissonia ovata).* Growing close to the ground, with seed that forms uniquely close to the base of the flowering stalk, this summer-deciduous perennial can thrive even in a mowed lawn. I am always happy to point out to clients its presence in their lawn; which may attest to the activities of our valued and threatened native ants, necessary to the spread of many native species.

Native seed-harvester ants can affect the spread of seed-bearing species in two quite distinct ways. First, they may consume quantities of common seeds, thus assuring that the seeds of less common species have greater opportunities to germinate and grow. Second, they can move seeds away from the parent plants without damaging them, relocating them in fertile rings of debris. So when I see new colonies of sun cups, which sets seed in ovaries very close to the ground and therefore cannot eject seeds a distance away, I hypothesize that native seed-harvester ants may be moving the seed. I hope that, unknown to me, some of the original ant life of our area continues, because the native ants of California are now in great danger.

The tiny, smelly ants that frequent your kitchen, particularly during the rainy season, are not native here but are transplants from Argentina. Argentine ants *(Linepithema humile)* are now the most common ant species in California. Thought by some to make one large, cooperative supercolony throughout the San Francisco Bay Area, these invaders get along with each other all too well. Native ants tend to waste time and energy fighting among each other rather than outfighting and outharvesting the visitors from Argentina. One possible control for Argentine ants involves a chemical that would cause them to perceive each other as members of a different, unrelated group. Disrupting their cooperative behavior could slow down their spread, which, in one study, occurred at the stunning rate of 32 acres in one year.

By disrupting the ways seeds are dispersed and consumed by native ants, Argentine ants are having a huge impact on California vegetation. California horned lizards *(Phyrnosoma coronatum frontale),* exciting denizens of the Central Valley and other arid parts of the state, are also being affected by the spread of Argentine ants. Now considered to be a species of special concern by the California Department of Fish and Game, these reptiles are dwindling in numbers, partly because of their inflexible preference for native ants. Argentine ants are not palatable to them, which, given their noxious smell in the kitchen, is understandable. (I have become so sensitized to this odor that I can detect the presence of a single ant in a five-pound bag of sugar.) California horned lizards love seed-harvester ants, and the rapid decline of our native ants eliminates a significant portion of their diet. Having followed this intriguing reptile through the wildflower fields of the San Luis Refuge in Merced County, I contemplate its disappearance with sorrow.

Like the European honeybee as compared to the native bee, the foreign insect forages for longer periods during the day than does the indigenous species. Argentine ants like disturbed soil and the wet conditions created by irrigation systems and sprinklers, reflecting the mesic, subtropical origins of this species. The gardener who eschews out-of-season watering can limit one factor in the spread of this destructive species. Though seed is spread by birds, wind, gophers, and flood, native ants are a major conduit, particularly of large seeds, and may be responsible for many of the surprising appearances of plants where you did not put them.

With the western scrub-jay to plant our forests, with California quail to prepare seed beds and plan sequence of blooms, with black-tailed deer for pruning and fertilizing, and with native ants to spread the seed of sun cups; I can at times feel almost unnecessary.

By helping with pruning, clearing land, planning sequence of bloom, spreading seed around, and fertilizing, our animal assistants give us time to make on-site observations, to form hypotheses to be tested right in the garden, and to search wilder but similar areas for clues and ideas. And also for sitting there having no thoughts at all, just being part of the hum and buzz, which leisure time is a gift of the quail, deer, ants, and, of course, the jays, partners in the game.

25 Celebrations from the Native Plant Palette

Summer weddings bring seed requests. Our wildflower seeds are asked to perform a variety of functions for the nuptials, from serving as wedding favors (frequently packaged in customized seed packets commemorating the bride and groom) to being flung at the happy couple in the place of rice. We are careful to explain that, while a noble gesture that honors our native flora, flinging a handful of seeds at the happy couple as they race to the car down the concrete steps into the asphalt of a church parking lot is not likely to result in a magnificent wildflower field at some future date. All we can know is that a variety of wildflower seeds will give the local bird populations more reason to rejoice than processed white rice. It can't be a bad way to begin married life, with the blessings of the birds, but it shouldn't be taken as an ominous sign when flower fields don't follow.

In a similar vein, teachers frequently ask me to recommend wildflowers that could be sown on Earth Day, in April, at a time when most wildflowers have long since germinated. It might be a truer celebration of Earth Day to sow in fall and winter, let seedlings grow with the winter rains, and teach something real about the ecology of California. Even with the best care, a spring planting has little chance of blooming before the schoolchildren have left for the summer, and even then, its blossoms, though still beautiful, will be briefer in their tenure. Sown at the right time, in preparation for Earth Day, the wildflowers will then be ready for the children to pick, lie in, draw, name, collect seed from, eat, and twine into flower crowns for their esteemed elders, the teachers.

I've also been asked to plan fall or winter sowings of those particular wildflower species whose colors will match the bridesmaids' dresses. Specific needs include bloom in a certain month, on a certain day, even at a certain time of day. The reliable clarkias come into their own for this job, since the summer months bring together both weddings and the flowering of many *Clarkia* species, and since shades of pink, which clarkias supply in spades, are often required. With so much riding on nature's performance, the hedging of bets, in this case, a sequence of sowing times and also both direct sowing and sowing in flats to be transplanted into four-inch pots, then later into the garden, eases anxiety.

I often think, in these summer months, that should I be asked to make floral arrangements for such festive occasions, I would go to the coastal hills. Passing a stand of creambush *(Holodiscus discolor)* in full bloom, I imagine a glass vase three feet tall holding these fragrant blossoms, whose graceful arrangement on the stalks lends itself to elegant display. The fuzzy panicles have a faint touch of yellow, the blossoms slightly past peak shading into elegant beige. Their scent, detectable from a distance, is lighter than the fragrance of lilac.

At the paint store I learn that there are four hundred different shades of white available in interior and exterior paint. Along the road on one June day, I see perhaps thirty distinctly different whites. Here are a few, all components of the bouquets I envision. The flowers of California buckeye *(Aesculus californica)* have a pink tinge, whereas the elderberries *(Sambucus racemosa)* that sometimes grow next to them are creamy. Like creambush, they have an enticing smell. Yarrow *(Achillea millefolium)* adds blue to an icy white, cow parsnip *(Heracleum lanatum)* echoes the cream of elderberry, wild cucumber is Navajo white. Thimbleberry *(Rubus parviflorus)* is pure white, like another June bloomer, ninebark *(Physocarpus capitatus)*. From the foothills, mountain dogwood blossoms also return to the pure side of white.

I imagine these sumptuous whites in an enormous vase next to the cake. The silky white petals of wild clematis *(Clematis ligusticifolia* or *lasiantha)* trail and droop onto the linen tablecloth. June, the bridal month, shows all that is contained in white.

26 Fog Flower

Four Guidelines for Backyard Restoration Gardeners

We are deep in the dense fogs of August. Outside my study window a multitude of sunny yellow disks float disembodied in the pervasive gray. Their scanty stems and petioles are scarcely visible, so that the blossoms themselves, with no visible means of support, seem suspended in midair. To say they are a cheering sight doesn't begin to express the impact of elegant madia *(Madia elegans)* blooming in the days of fog.

Its lemony yellow petals are fully open only at the beginning or the end of the day, or when, as happens regularly in August, we are enshrouded in fog. Now I say to garden visitors on a sunny day, "Oh, if only you had been here before the sun came out." It is truly wonderful to wander in a tarweed forest, inhaling its complex soapy fragrances that smell in a way like kerosene, were the smell of kerosene entirely pleasant.

This ebullience results from a previous year's container planting of this annual wildflower, also known as "common tarweed," so it is both elegant and common, a winning combination. When the plants flowered, went to seed, and died, they dropped seed all around the container, and five years later elegant madia is everywhere. It's a major feature of the front yard both when in bloom and later, when its drying leaves and stems form a fine, graceful haze, till I submit to fire-safety considerations that require the removal of large masses of dead vegetation. Before that, though, I enjoy at least a month of its graceful decline, providing, all in all, a most creditable performance for an annual. This surprise is but one of the many guidelines that "fog flower," as I call it, provides for the restoration gardener.

Guideline no. 1. Expand your aesthetic sense to appreciate the subtleties of annuals in their post-seeding decline. Grow them in masses to make their visual attributes at this later stage more striking.

Two genera, *Madia* and *Hemizonia,* contain the species most commonly known as the tarweeds or, as they are less frequently called, tarplants. All in all, they include twenty-one madia and nineteen hemizonia species. Tarweeds can be annual or perennial, white or yellow or yellow with orange blotches, short, tall, slender, stout. Some are rare; some are common. They can open in the morning and close at noon, or close in the morning and open at noon. They are generally sticky, they all have two kinds of seeds—ray and disk—and they are usually aromatic. Their odors are variously described as fragrant, mildly scented, strongly scented, ill-scented, and strongly ill-scented.

I am a fan of the clean, pungent odor of most tarweeds. For me, it is a good smell, a late-summer-in-California smell, fresh and bracing. Hayfield tarweed *(Hemizonia congesta)* has a sharp, interesting odor and in some years turns our coastal fields to gold in midsummer. Another species whose fragrance I hope someday to experience is *Madia citriodora,* or lemon-scented tarweed, which can be found in high places in the North Coast Range as well as in the Sierra Nevada.

As with the genus *Clarkia,* tarweeds are a good way to locate yourself in California. One home owner in Los Gatos, looking for a ground cover for the slopes around her new house, went to a nearby hillside to collect seed of *Hemizonia congesta* subsp. *luzulifolia,* also called hayfield tarweed. After two years of removing invasive thistles and persisting with direct seeding of hayfield tarweed, she has a self-renewing cover that is green early in spring and provides soft white blossoms in midsummer. The pleasant and uniquely Californian odor drifts, on a sunny day, up to her shady porch. Her creative use of tarweed earns this gardener an honorary membership in the Tarweed Appreciation Society.

Another customer queried us about our listing of *Madia elegans:* "Is this the classic, 'your-teacher-will-hate-you-but-appreciates-the-thought,' sticky, pungent tarweed all kids who walked to bus stops in the country (and came home with black sticky marks all over their hands) know and love?" It may be *Madia elegans* she remembers, or one of a slew of other madias and hemizonias found blooming in the hot, dry reaches of California. "If so," she adds, "I'd love to get a sufficient amount to thoroughly plant my surrounding area and the drier portion of my yard in Oroville, California." Another member of the Tarweed Appreciation Society.

The way they dry and the way they die is a notable garden feature of madias. In October the seeds of elegant madia rest intriguingly in the dried flower heads, the stalks forming a delicate tracery of gray. How annuals go to seed, the way they shatter, how their crisp stems hold on through the fall, to finally be battered into submission by an early downpour, these are noteworthy gardening events, more easily caught by the eye when madias have expanded their territory to a relatively large area, about one hundred square feet. The pollinators prefer such a patch to scattered populations also.

Yellow tarweeds have their own slant on yellow. Add just one drop of green to pure lemon yellow, a mere hint, but enough to keep your eye returning in tarweed-yellow addiction. In the fog, coming around the corner of the house, I am repeatedly stunned by its brightness against the gray siding and the gray day. When complimented on this unintended effect, I modestly accept the praise.

It takes discipline not to blurt out that this beautiful tarweed occurrence was unintentional. After all, I did not plant all these tarweeds, but I did create the opportunity, in this case the bare soil, the weed-free ground, for them to express themselves in this pleasant way. I planted the first population and readied the way for the second.

Guideline no. 2. Be open to the results of happenstance—to elegant tarweed providing months of garden interest, to hayfield tarweed pinning a new house to the hillside—and though they were not the result of your careful, knowledgeable planning but generous gifts, graciously accept praise for them.

This fall I had another opportunity to practice my second gardening guideline. The scenario involved a planting of the evergreen subshrub Catalina currant *(Ribes viburnifolium)* with snowberry *(Symphoricarpos mollis).* When the leaves of the deciduous snowberry fell, leaving showy white berries hanging from gawky bare stalks, the attractive, dark-green, glossy leaves of Catalina currant rose up behind them, as though supporting its dormant friend. It looked like a new species with two desirable garden features: pretty white berries in fall and handsome evergreen leaves. When complimented on this effect, I took all due credit.

Guideline no. 3. Somehow get an entomologist to your garden. Have a resident entomologist, who lives on-site and is there to explicate insect events. Or else encourage your daughter to become one or marry one. If that isn't possible, rent an entomologist by the hour.

I share my love of tarweed with a plenitude of insects. Once I collected seed from the *Hemizonia congesta* mentioned above, the fresh-smelling hayfield tarweed. In doing so, I disrupted bug paradise, a heaven of insect life going on in those seed heads. The collecting basket into which I beat the seeds was soon teeming with a diverse mass of insects madly trying to escape, now that they had been dislodged from their snug seed homes.

These pungent, late summer–blooming native wildflowers are reputed to have been an important food source for the native people. In Grace Carpenter Hudson's painting *The Tarweed Gatherer,* held in the collection of the Grace

Hudson Museum in Ukiah, California, a member of the Pomo tribe is depicted employing a seed beater to gather tarweed seed. One writer referred to tarweed as "wild wheat," saying that the Indian's autumn begins when "the *le-mo-lo sap-O-lil* (wild wheat or tarweed) had all been gathered and winnowed and the whole countryside could now be baptized with fire." I am told that jars of tarweed seeds have been found in anthropological collections. In spite of its apparent paucity of seed, it seems that the elegant tarweed is here to stay, reseeding all around the wooden container and moving into the raspberry patch. Still, I will take no chances.

Guideline no. 4. When growing special plants in the garden, the small things, hedge your bets. Plant them in several different places, with different kinds of care. Take no chances with the precious ones, like the common and elegant tarweed. Honor chance and the sway of unknown factors.

When we are lucky enough to get native clover seed, we always grow some in large containers protected by copper strips from snails and inside a deer fence. If we plant them in the ground, we protect them from gophers, snails, deer, and weeds. When growing rare clarkias, we try different kinds of soil, finding that frequently the rarer species like extremely sharp drainage and lean soil, eschewing bat guano and manure. I coddle two flats in two different greenhouses with subtly different conditions and place another flat directly outside.

To summarize, here are four guidelines: grow in a mass to turn subtle, seasonal features into design elements; take credit for unintended successes; have access to an entomologist; and hedge your bets. These are mine, emerging from work in the backyard restoration garden. Take notice, as the years pass, of your own.

27 Scale Is All

Large Shrubs and Small Trees

One sunny Sunday in July I am drowsing under a blue blossom ceanothus. Flocks of bushtits are gleaning and chirping away in the foliage above me. The blue blossom ceanothus makes a graceful umbrella over my head, one I have enjoyed for many years now, ever since I planted it to replace a cedar that fell in a storm. When the cedar fell, I needed quick shade, something fast-growing to take its place. Unlike the forest giants frequently planted for shade, windbreaks, and privacy hedges, *Ceanothus thyrsiflorus* reached its ultimate height of fifteen feet within three to four years and stayed there. In the spring, its sky blue blossoms are sweetly fragrant and humming with pollinators. Its shiny blue black seeds draw all kinds of birds and a myriad of insects hover around it at all times, making it a center of life, a great place to nap, to read, to pursue all kinds of daily activities. It is just the right height to provide shade for a person sitting under it, and it does not threaten the office.

I am almost asleep when an all-too-familiar noise shatters the peaceful afternoon, obliterating the delicate chatter of bushtits. One chain saw to the right and one chain saw to the left, the weekend removal of the forest giants has begun. The single greatest expense for many gardeners can be the removal of too-large trees near their homes. Up and down California, the same trio of tree species requires the same endless drone and the same use of gasoline: Monterey pine, Monterey cypress, and blue gum eucalyptus. Things might quiet down if we were to use only those plants that reach an ultimate height appropriate to our homes and to our utility lines, that do not demand constant pruning or removal, and that do not cause power failures during our many and increasingly frequent storms. Every part of California has good choices, both evergreen and deciduous, from this category whose praises I sing, the native large shrubs and small trees.

Tree removal is costly. Several tree surgeons in my neighborhood enjoyed extended Hawaiian surfing vacations paid for with the money I spent removing eleven Monterey pines from the immediate and dangerous vicinity of my house. Equally expensive in the long run is the torturous habit of pruning to control height, called "topping." An organization from Seattle called PlantAmnesty has taken on the cause of eliminating tree topping from the pruning repertoire. Its founder, Cass Turnbull, says "the bottom line is that poor pruning results in decayed trees. A big limb can crash down due to a bad cut made 20 years ago. . . . Pruning doesn't prevent hazardous trees—it creates them."

Cutting the upwardly mobile parts of a tree sets it up for a number of problems, including the creation of internal columns of rotting wood, the production of weakly attached and unsightly regrowth, and a generally hideous appearance. Much easier is to avoid inappropriately sized trees in the first place, and where they already exist, to bite the bullet and opt for removal, facing a problem that will only become more expensive and dangerous through time.

In a typical scenario, a home owner in Cupertino removes a massive Monterey pine from his small yard. Paying the extra money for stump grinding, he is able to replant the area immediately. Saving the trunk, he has it milled into bench material. He replants with hollyleaf cherry *(Prunus ilicifolia),* a beautiful small tree that can be pruned to look like a small oak or left bushy for a privacy screen. Evergreen, with glossy, medium green leaves and fragrant white flowers beloved by bees in the spring, it was once a distinctive component of many parts of California. Important to indigenous peoples for the kernels of its large purple black fruits in midsummer, it was given the common name of islay, derived from the Salinan word *slay.* Place-names honoring this species include Islais Creek, Islay Bridge, Islay Hill, and Spanish grant deeds like El Arroyo de los Yslai. Within three years, this shrub grew to the height of the fence; within five, it's reached its ultimate height of ten feet tall, though in riparian zones or heavier valley soils it might reach twenty-five feet. Threatening no power lines or houses, requiring no expensive pruning, it provides shade, evergreen privacy, attractive flowers, foliage and fruit. It is good wildlife habitat, drought-tolerant, adaptable, and attractive through the seasons.

The single greatest expense incurred in urban gardens is removal of excess plant material, whether it is pruning of forest giants or removal of out-of-scale trees, shrubs, or perennials. The single greatest temptation for the new gardener is to make a new landscape look established as quickly as possible through overplanting, without taking into account the ultimate size of the shrubs and trees.

Confronted with a blank slate, the impulse is to fill it. It's difficult to imagine the ultimate height and width of these plants, particularly since height and width charts sometimes present wide possible ranges. A given plant, for example, salal *(Gaultheria shallon),* is listed as growing from two to ten feet tall. Not much guidance there for the gardener. Each site being different, it may take ten years in one case, or twenty in another, before the ultimate height is reached. When the design is installed, the client understandably wants to have the impression that mature plant materials are in place. This requires the use of large-size plant material in a situation where money could be saved if smaller containers were purchased, which would also result in healthier and longer-lived specimens.

I know urban home owners who never get around to actual gardening, because all their resources go to plant removal. The expense of buying plants and putting them into the ground pales by comparison. Money spent in pruning or removal of trees and shrubs that are too close to each other, or too close to fences, or too close to houses, includes the cost of hauling these materials to municipal dumps. The smaller the garden, the trickier removal is, and the more frequently it is required. The smaller the garden, the more noticeable and painful is each removal. In urban situations, every tree is so precious, to both human and animal life, that its absence is deeply felt. So planting correctly in the first place is a blessing for yourself and those who come after. Here's a litany of possibilities from up and down California: hollyleaf cherry *(Prunus illicifolia)*, Catalina cherry *(Prunus illicifolia* subsp. *lyonii)*, Parry manzanita *(Arctostaphylos manzanita)*, bigberry manzanita *(Arctostaphylos glauca)*, tree ceanothus (*Ceanothus* spp.), toyon *(Heteromeles arbutifolia)*, coffeeberry *(Rhamnus californica)*, Pacific wax myrtle *(Myrica californica)*, hazelnut *(Corylus cornuta* var. *californica)*, elderberry *(Sambucus racemosa)*, tree anemone *(Carpenteria californica)*, California fremontia *(Fremontodendron californicum)*, and vine maple *(Acer circinatum)*.

Employing plants that know their limits, we can once again enjoy the light chatter of bushtits, the rustling of towhees, and the calls of jay and quail on sleepy Sundays.

The Fifth Season

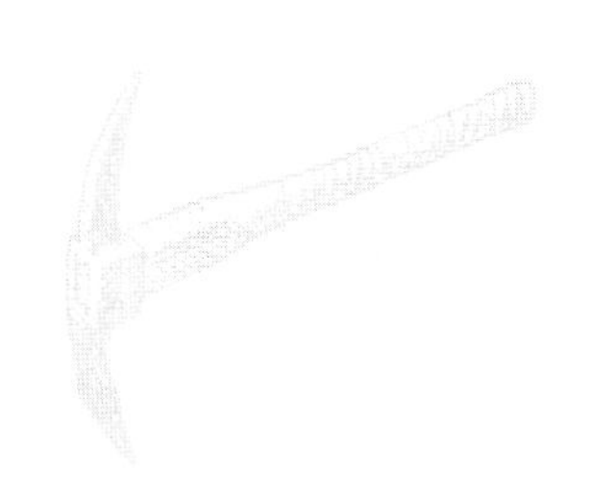

28 The Quiet Time

We need a name for this ineffably tranquil time of the year, when the effects of four to six months with no rain are felt in the landscape and in our gardens. With no daily expectation of rain, plants, as well as some animals and people, lock onto a holding pattern, a still dormancy. Walking through the dry woods is a crunchy experience; sitting in sun-baked chaparral, the small rustlings of lizards alternate with a solemn stillness, a consequence of this long time of aridity. This unique moment occurs only in the Mediterranean climates experienced by 2 percent of the world's land masses.

Understanding the constraints and opportunities of this particular season might be easier if it were distinguished by appropriate vocabulary. David Rains Wallace refers to California's dry time as "the fifth season." He suggests that we hold festivals to celebrate the moment in the yearly round when life seems to pause and the breath is held before the nourishing onslaught of the fall and winter rains. I further define this excellent term to refer to the time between summer and fall, when, with no expectation of rain, we come most stringently to terms with what it means to be Californian. Shady places, within which the contrast between light and dark is exquisitely defined, become almost painfully beautiful. Newly dropped leaf litter crackles underfoot. In those moments between noisy quail calls and the shuffling in leaf litter of brown thrashers and towhees, a quiet comes, a stillness born of aridity, felt intensely on sunny afternoons on golden hillsides.

Days shortening, birds on the move, shadows lengthening, plants going deeper into or in some cases coming out of summer dormancy. It's hard to remember now the ecstatic discoveries of spring, that frenetic energy, all the projects I began then, and the summer dreams of all the hard-to-find seeds we might include in our listings and seed-growing projects, somewhat difficult to reconcile with this moderated state of low-key activity. It's hard to remember moisture. Rain? What is that? Though the gardener can employ strategies to ensure some bloom even in the fifth season, for example by planting wildflowers in spring and providing artificial irrigation, I like to accept the fifth season as a time of little, or no, bloom, relying on silvers, golds, ecrus, beiges, and browns, as well as on the shapes of things.

The fifth season can be a time of planning strategy for fall. Wildflower beds can be prepared for the fall sowing, and design decisions and plant selection in nurseries can begin. In my town, we know that adding wood chips to raise paths in areas with poor drainage or high water tables will ensure a dry journey from car to house or through the garden. So the fifth season can be a good time to clean up, prune, and clarify paths and trails, getting ready for fall, the high-activity time in the garden with which this seasonal journey began. But the quietude calls, and frequently, I respond.

29 The Berry Harvest

One afternoon during the fifth season a friend visits, bringing me up to date on her life. While she talks, I am cleaning seeds, focusing on late summer berries, such as pink flowering currant, salal, evergreen huckleberry, hairy honeysuckle, elderberry, salmonberries, coffeeberry, and wild grape. At the same time, I set aside some huckleberries, elderberries, and wild grape to dry for my own wild raisins.

As I work, I acknowledge this bounty and the longtime importance of berries in human life. I recall what Lantz and Turner said, in their discussion of phenology, "Traditionally, berries were one of the most important food resources, and served as an essential winter foodstuff. Berries were also extremely important in trade, and as a food gift item in potlatch ceremonies." These dark purple, red, and deep blue fruits are rich in anthocyanins, phytochemicals replete with what may be a range of life-protecting effects.

I put the berries with some water in the food processor and turn it on briefly, till the berries have been macerated. Then I pour the whole mass into a large stainless steel bowl, add more water, and leave it for a few minutes. Separation of fruit from seed will now take place with no help from me, as the heavy seeds settle in the bottom and the fleshy fruit rises to the top. The dark berries tint the water rich purple. My friend interrupts herself—"That's a gorgeous color."

While she continues with her story, I tilt the bowl so that the water and all the skins and flesh that have risen to the top can pour out of the bowl and into a second bowl, which is a kind of insurance against losing seed with enough flesh still attached to be light enough to slide out of the bowl with the debris. As I get closer and closer to the end of the water pouring, I am in some suspense to see if I will find heavy, insect-free seeds waiting for me at the bottom of the bowl. When seeds have been invaded by seed weevils or fungus and are no longer viable, they usually float to the surface, being too light to sink to the bottom of the bowl.

Just before the heavy seeds start to slide out of the bowl, just before the water is all gone, I stop tilting the bowl, then pour in more water, and rub the seeds between my fingers to detach any remaining fleshy part. Ceasing activity, I let the water still itself one more time, so that once again heavy, healthy seeds can sink to the bottom, while the debris, newly separated from the seeds, can float to the surface. Again I pour out the water, which carries the light debris with it.

I may repeat this part two or three more times; each time there is less debris and the seeds are cleaner, until finally I pour the entire contents into a strainer. In spite of all these efforts, there may still be some debris attached to the seeds. Tapping the contents of the strainer out onto a china plate, I place it in the greenhouse.

With one part of my mind on my friend's story, the other part is engaged with this physical experience that is taking me deeper into the properties of the plants. The consistency of the berries, the size, color, and shape of the seeds, the difficulty or ease with which they have been separated from the fleshy fruit, these details occupy me. The discovery of their final feel and look tomorrow, when they have dried in the seed room, awaits me. In the morning this gloppy mess will have dried into tiny beige, brown, gray, black, or golden seeds, which I will rub between my fingers and along the surface of the plate to separate, reveling in the sound the friction produces, in the way they look, and in how they feel to my fingers, an indescribable pleasure, an intimate, particular knowledge.

While cleaning them, and while getting them from one container into another, I have consciously and unconsciously savored the feel and the play of seeds. In the course of my work I engage with the movement of seeds from one place to another, pouring seeds from bags into jars and from jars into bowls and from bowls into packets and from packets into flats or into the ground. I pour them from baskets into jars and from jars into baskets. To facilitate the pouring, I use a funnel of metal, plastic, or paper as well as my own cupped hands, my preferred method. For seeds that are not embedded within fleshy fruits, hands, rolling pins, and even feet will separate the chaff from the seed. Winnowing to lift off the chaff can occur outdoors, when gentle breezes remove the light dry material, leaving the heavier seeds at the bottom of the winnowing basket, or in the absence of a breeze, in front of a fan.

My friend asks, "Isn't that tedious work?" Somehow, it's not. For many thousands of years, the cleaning of seed foods has been a part of the human experience. I relish my tiny part in it. Usually the seeds will be measured and put in small packets with the appropriate label affixed, but sometimes, if I know that these are food, I roast them quickly in a cast-iron pan, grind them a bit, and eat. Each species has a different taste, a different savor.

Edith Van Allen Murphey, according to her biographer Skee Hamann, described Lucy Young, of the Lasski-Wailaki tribe, processing acorns when she was a hundred years old, "so patterned to succor, so conscious of need of nurture in famine times past that she processed medicine plants and prepared acorn meal even after she was blinded by cataracts." Among her last links to life were the texture of the acorn, its heft and weight, the way the meat sits inside the shell, and the way it can be removed, dried, ground, and leached. The acorn woodpecker and the western scrub-jay are not the only creatures shaped by acorn. Long-time human inhabitants of California were also shaped by the physical properties of the seed of oaks and many other plants, which determined not only their nutritional intake but also the work of their hands, their sense of what was delicious, their feel for and their take on the world. Today, some of these qualities have shaped my afternoon. Stopping her story, my friend asks to help clean the berry harvest, so I pass her a bowl of dried currants and let her pick them apart with her fingers, spilling the seeds out onto a dish. Between us now, there is silence.

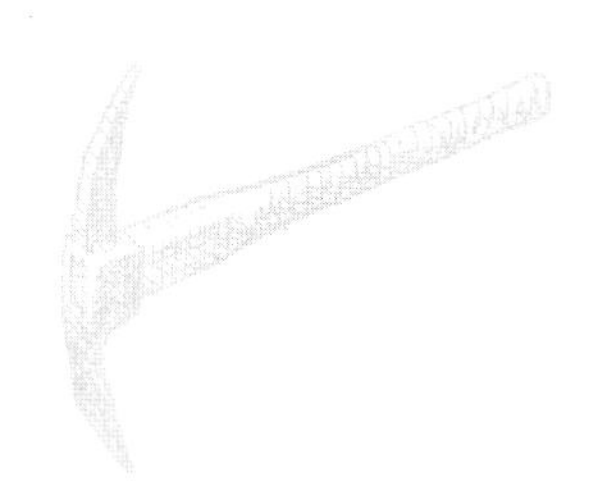

30 Paths and Trails

Making Our Way through the Restoration Garden

> The trails were sacred. "Just the same as people," one old man said. It was wrong to step out of them without some good reason. There were established resting places and places of offering along them where a prayer was made.
>
> Pliny Earle Goddard, *Life and Culture of the Hupa,* 1903

With little else happening in the garden during the fifth season, it is a good time to work on paths and trails. Paths are a significant element in the garden; organizing the space, they control our experience of it. They are the threads our eyes follow. They determine what we brush against, when we walk freely, and whether we walk upright or stoop to go under a coffeeberry branch. Determining the foot's experience, whether of soft, silent, and absorbent materials, or of hard, crunchy, gravelly, and pebbly substances, paths and trails affect the quality of our movement through the garden.

The options for paths are dizzying in their variety. From elaborate patterns of brick or stone to the country lane, from the formal allée to the mulched path, ways to get through gardens fill entire coffee-table books. I myself could write a chapter about one of my favorite ways to move through gardens, a way that is inexpensive, low-maintenance, and singularly appropriate for the backyard restoration garden. Relishing the look and feel of well-used but not overused dirt paths, I recommend dirt or leaf litter paths and trails as an appropriate way to make our way through the backyard restoration garden.

Because of our complex geology, some parts of California may have as many as nine different soil types within one square mile. Dirt trails celebrate this diversity by exposing it to the eye and the foot. The basic connection between the human foot and the particular soil it treads, whether gray, beige, orange, reddish, brown, yellow, or black, is one the restoration garden may promote. With nothing to impede the connection between foot and earth, dirt trails give you the chance to directly experience your soil, whether it is bedrock, clay, gravel, or sandy loam. A dirt path says, "This is the soil out of which this particular garden is made, the soil that grows these plants. See it, feel it, walk on it."

When I see a game trail through the chaparral, my heart seems to beat faster. It is as though the trail calls to the foot. An artfully composed stone or brick path has an entirely different feel, calls attention to itself as a separate creation, set off from and apart from the land. It can be a delight, part of the more formal approach to the house—I do not seek to eliminate all nondirt paths, only to elevate the dirt path to its rightful place in garden design.

Made of the soil that supports the plants, the dirt trail that winds through the garden not only subtly reminds us of the wild but possesses certain practical advantages as well. A hardscape path of brick, stone, or cement may easily absorb a greater proportion of the money, materials, and labor spent on a garden than do the plants. Saving money and time on paths frees those resources for other things, such as the preservation of small things, or gorgeous containers to call attention to and protect herbaceous species, or realistic budgets for weed control without chemicals. As well as being the least expensive alternative, dirt paths allow for easy change and thus are a good way to experiment with possible layouts. As plants grow and gardens develop, paths can be changed with impunity. They also change themselves. They can disappear through time, made impassable by the growth of trees and shrubs, or by the seeding of herbaceous species. There are so many ways to carve up space, each with their own suite of advantages and possibilities. Change can renew your experience of the space and also be good for the plants and animals. Quail are thrilled when we make changes, the first on-site to explore new opportunities for scavenging and dust bathing.

Dirt trails make good seed beds, and we frequently find many precious seedlings growing in the trails in late winter, which we repot and move elsewhere. Or, when many small species have begun to seed into this good seed bed, we carve a path somewhere else, letting the old path fill in. I create a new view of the thing, cut a coyote bush to the ground for its own good, choose the larger of two oak seedlings and remove the other, reroute to open a new prospect or view.

Trail makers in our national parks routinely close old switchbacks and establish new ones. On steep slopes, letting old trails reestablish their cover while routing new switchbacks allows the land to heal. These techniques can be used in the garden too.

Dirt or mulched paths can be labor-saving. Though you may opt for ever-renewed fancy raked patterns, those are optional; dirt paths don't need to be swept. The soft fruits that fall on them will be absorbed into the earth. Weeding between bricks and pavers can be a maintenance headache, one of the few situations where I have been tempted to use herbicides.

The backyard restoration gardener can play with ways to get from one place to another. One strategy is to have a range of styles mirroring the range of maintenance strategies. At the outer limits of the garden, where maintenance strategies may be more subtle and possibly less frequent, the dirt or mulched path can be used. At a certain point, closer to the house, the route may change to hardscaping. By the time you reach the door, the dirt will be scraped off your shoes rather than tracked into the house, one of the practical reasons for hardscape pathways.

In the restoration garden, establish a trail using wood chips of one kind or another and then allow the chips to decay without replacement, or remove the remaining chips and leave a bare dirt trail for awhile. Leaf litter also makes wonderful path material. When the trail leads through different plant associations, each with its own manufacturers of leaf litter, the walker may experience the crunch and texture specific to bay leaf litter, oak leaf litter, or Pacific wax myrtle leaf litter, deepening the sensory experience of the garden.

Materials for unique paths can come right from the garden itself. In one garden edged by a wetland, sedges (*Carex* spp.) grow abundantly. Annual clipping to the ground allowed resprouting with fresh new growth in the spring. This creative path maker took the resulting sedge clippings, cut to four inches long, and laid them lengthwise along the path. As the clippings dried, they turned from rich green to pale green to brown and beige, eventually becoming almost inconspicuous, but continuing to lend a bounce to the step. Sedges growing on the side of the path, sedge clippings under your feet, and sedges all around; the theme of this garden was sedges.

THE ATTENTION REQUIRED BY TRAILS

Near the beach where I live is a beautiful hillside, half of which is perfect, intact coastal scrub, and the other half of which is perfect, intact coastal prairie. Because the hard-packed old trail up the steep and sandy hillside had begun to erode, it was recently rerouted by a public agency. The agency carefully switchbacked a new place through the scrub and conscientiously closed the old trail. All was well till the following spring, when weeds found the soft new path. Wild lettuce, thistles of all kinds, and French broom appeared where they had not been before. From the path they began to make inroads into the previously intact scrub and prairie, posing a clear threat to one of our few, if not our only, relict stands of scrub and prairie.

The local land trust noticed what was happening and organized a work party, which was attended by a representative from the governing agency. The new path in itself was a good thing, but the opening it provided for weeds to make inroads where previously they had not been was a good example of what happens with all ways of passage, from trails to freeways.

It matters what happens to the roadsides. Though in some situations, grazing inside a fenced area controls weeds and protects native species, and the ungrazed roadsides outside the fence accommodate weedy intruders, in many other places throughout the state, steep, rocky roadsides outside the fences are some of the last refuges for natives, providing important clues to what used to be here. In either case, as the weeds' staging ground into our fields and wildlands, or as the last holdouts, roadsides matter. We need road warriors to protect them.

TRAIL MAKERS

One of my favorite trail makers is a small herd of wild horses that was transplanted twenty years ago to the Mendocino ranch where my forest restoration takes place. After an unsuccessful attempt to capture and remove the herd to yet another home, nine horses remain. Because they are few, nonreproducing, and the land they range large, I consider that some of their impact may be beneficial. One of those benefits is a series of elaborate trails through the grasses.

It is comfortable following the horse trails. They are about a foot and a half wide, with sharply defined edges and beautifully pounded soil, creating a place to walk that is free from poison oak, ticks, and rattlesnakes. These attributes of openness and visibility were traditionally valued in California, where people used the same trails for thousands of years. Trails were regarded as conscious beings with the power to confer blessings if treated properly. In 1927 Edward Sapir recorded Hoopa Sam Brown's trail prayer: "[You] lie here still. Let me go about without criticism. Let me grow old doing things this way. Let me keep coming back along this trail only with something good. Let me live happily."

Effie McAbee Hulbert, who grew up in the Anderson Valley in close association with a Pomo group that lived on her family's ranch, describes the old-time trails in an autobiographical novel through her fictitious character Muchacha:

> [She] wondered what their life had been elsewhere; what indignities the mother of Sarah-Steena might have endured, as she went on along the hard-as-cement trail along the river to complete Grandmother's errand at Uncle Van's. All Indian trails were this way, even the ones in the sand, packed hard; and she thought of the white man's trail, rough and dusty, even on hardest ground, pitted by heavy boots and shoes.

Since use of the trail helps me to maintain it, and since I want my garden to be viewed and appreciated, I encourage friends and neighbors to take a shortcut through my garden as they walk around town. To this end, I have laid out a way to move from one end on one road to the other end on another road. This route begins at one gate with oak leaf litter underfoot, moves through chipped coyote bush underfoot, then over what I call California's Exotic Mulch Mix, a combination of local, non-native trees or misplanted native trees that require removal. That would include, of course, the ubiquitous eucalyptus, Monterey pine, and Monterey cypress, as well as acacia. This mix, with other woody components, is widely available throughout the state from city, county, and private tree crews, and though perhaps not ideal, we have had good success using it.

In the main seating area, where a steady state of chipdom is desired, the garden visitor traverses Douglas fir bark chips, a frequently used material widely available from nurseries and landscape materials yards. Douglas fir bark, though a by-product, is not my favorite path or mulching material; the chips are too large underfoot to be comfortable, too uniform to look natural, and costly to deliver. They are, however, relatively long-lasting and are useful where dirt trails have the potential to become muddy ruts in the wet season.

Because of misplanted trees in California, there is a constant local cutting and chipping of arboreal materials in most cities. These chips can be useful, and the price, free to modest, is usually most acceptable. They have an appropriately uneven look, and they represent the removal of non-native trees, the by-product being the mulching of native trees with the demise of non-natives.

When you use dirt paths and trails as a way to move through the space, every visitor helps to maintain them. Every visitor, including deer or elk, is a trail maker. The number of sandaled, moccasined or bare footsteps needed to create the kind of surface Effie Hulbert describes is unknown, but they represent the journeys of many ancestors. No wonder trails were thought to be sacred, so many feet followed by so many other feet. In modern times, as we continually bushwhack our way through life, it is restful to contemplate such a trail and to think of re-creating it on our own land. May your garden be such a well-traveled place.

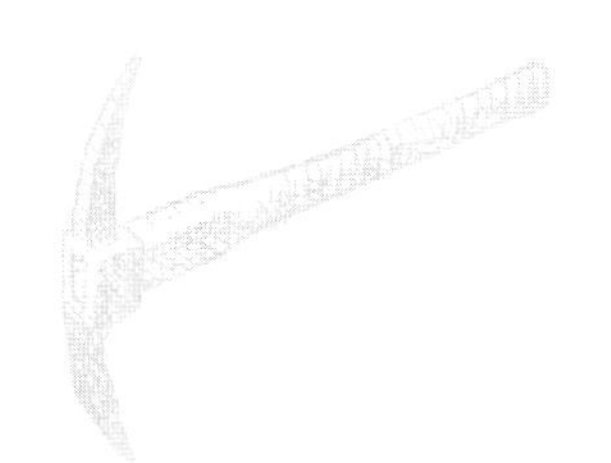

31 Everything's Here

The Late Summer Riches of the Pond Scum Production Area

More important for birds than providing bird feeders is providing water.

Geoffrey Geupel, director of the Terrestrial Ecology Division of PRBO Conservation Science, 2004

It's hard to have a pond.

Stephen W. Edwards, director of the East Bay Regional Parks Botanic Garden, 2004

On the coast in September and early October we usually enjoy Indian summer, our fifth season, the payoff for summer fog, while inland it boils. Now the pond comes into its own. Suddenly, butterflies sip from it and flit across it, damselflies and dragonflies emerge from it, water striders surge across the surface, and the pond is full of creatures, food for the migrating birds. Many baths, drinks, and meals are provided by this simple pond.

I made the pond before I had a model in mind, an act of faith. Lined with ferro-cement, it is about four feet deep and eight feet long and five feet wide. I had hoped to feel my way into the situation, avoiding as much complex pond technology as possible. Not inclined to deal with pumps, which kept clogging or falling over, or fountains, whose noise seemed too obtrusive for this garden, I settled for what seemed for a while to be a stagnant hole filled with pond scum. The success of my tule plants, including California bulrush *(Scirpus californicus),* a hollow-stemmed wetland plant with edges, or common tule *(Scirpus acutus)*, without edges, and my subsequent exploration of freshwater marshes gave me the model I was looking for once the pond was in, while the red-legged frog provided a few necessary guidelines.

ASK THE RED-LEGGED FROG

The California red-legged frog *(Rana aurora draytonii),* the largest native frog in the western United States, is in sharp decline in California. A predictable combination of factors, from habitat loss, pesticide runoff, climate change, exotic non-native predators or competitors, like the introduced bullfrog, as well as drought and grazing, has knocked out populations in fifteen of the forty-six counties where formerly they resided. Freshwater marshes everywhere are in decline, and the freshwater ponds that are part of many marshes seemed like a possible model for my pond.

Red-legged frogs like deep pools in still or slow-moving water, and successful populations are associated with overhanging willows and a fringe of cattails *(Typha latifolia)* or tules (*Scirpus* spp.). In summer, red-legged frogs move as far as one hundred feet away from water to estivate, the dry-season equivalent of hibernation, in small mammal burrows. Good estivation habitat can be provided by the gardener using any of the riparian species, such as willows, sedges, elderberries, hazels, sycamores, ninebark, and a host of others. If no small mammal burrows are available, leaf litter can provide a place for summer estivation. Many local ponds in this area are still jumping with tadpoles and frogs.

A ledge runs along the edge of my pond, on which are placed containers of bulrush and horsetail *(Equisetum arvense).* These frequently fall into the pond and allow the plants to grow there. Cement blocks in the middle of the pond hold smaller native aquatics in pots, like sedges, including *Carex barbarae,* Point Reyes checkerbloom *(Sidalcea calycosa rhizomata),* yellow pond lily *(Nuphar polysepalum),* and the fascinating arrowhead *(Sagittaria latifolia).* Also called wapato, or tule-potato, this erect winter-dormant perennial, with arrowhead-shaped leaves, has white-petaled flowers in late summer. A famous indigenous food source, wapato was traditionally harvested with the toes. Lewis and Clark enjoyed these nutritious tubers in Oregon, and they can also be found in slow or unmoving fresh water in Canada and eastern North America. Because wapato goes dormant, you will need to find a safe place for seemingly empty containers with good labels indicating that something valuable sleeps within.

REMOVE SOME, LEAVE SOME

Classic green pond scum appears when the rains stop. It is a prodigious grower, bright chartreuse, with long filamentous strands. When manually removed from the pond with a net and left in the sun, it dries into a whitish absorbent spongy substance, with potential use as sponges, diapers, and in paper making and other art projects. Conventional pond wisdom might have it that these algal blooms are a sign of ill health. My personal algal specialist, Eugenia McNaughton of the United States Environmental Protection Agency, doesn't necessarily agree. She finds beauty, fascination, and worth in pond scum and feels that it has an important place in even such a small, vague project as mine.

> To the unmagnified eye, algae are nothing more than a massing of color, greens, yellow-greens, blue-green, brown, more rarely red in a fresh water pond, lake or stream. They sit tight on rocks, flow in green strands along a pond or stream bottom or color the water. If you try to grab them, they prove to be wispy, losing their shape out of their environment. If you take a water sample, you can see the cells swimming around. Neither plants, nor animals, what are algae? Using sunlight to make sugars, they are the food at the bottom of the great chain of aquatic being. Without them, there would be no food for the herbivorous invertebrates and fish that are the food for larger invertebrates and fish, and all the aquatic-dependent wildlife, such as birds. Algae offer places for aquatic creatures to rest, to hide from predators, and to lay their eggs. If you look at them with a hand lens or a microscope, you discover a diversity of form and organization you could never have imagined living in a drop of pond water.

The kinds and quantities of pond algae and other water plants in my pond change through the seasons and are different every year. Sometimes in the middle of a rainy winter, delicate round pink units called azolla, or "water fern," appear, which, though not good for sponges, make a particularly exquisite mulch. Fish and shrimp relish it, and I intend to try it soon as an addition to soups and salads, for, again, as William Least Heat-Moon said of his culinary experiments with the native plants of Kansas, "a writer most fails when he loses nerve." Dried, it is said to be nutritious as well. To grow water plants in containers outside of the pond, I place them in larger, nondraining containers filled with water. To prevent these containers from becoming mosquito breeding sites, I transplant azolla from the pond, adding it to the water in the larger bucket; mosquito larvae cannot come up for air through its dense mat.

TULE CULTURE

When it is warm enough for dragonflies to lay eggs, it is time for me to get into the pond, where I fish for fallen sculptures and pots that have tipped over into the water. I harvest pond scum for mulch to dry for outdoor-use sponges, leaving some for dragonfly egg-laying activities and shade in the water. Also on this day, I separate tules, so that one large plant becomes many. Once I succumb to the cool water, the muck, the silky slime of pond scum, the predatory dragonflies and their larvae, the crunch and snap of tule stems as they are teased apart and set up in separate establishments, I become content and rest in the work.

This procedure, distinct from dividing other plants, involves being both brutal and delicate. Related to the bulrushes in which Moses took his infant journey, *Scirpus californicus* has long hollow stems that are easily broken, and if they break anywhere along their length, they must be removed at the base to avoid an unsightly stub. They require respect in the way they want to stay together and must be urged gently into becoming separate but equal, because we know it will be good for them, and good for the frogs, if they become many. In this small way, I commemorate what some call "tule culture," the intense interaction with tule for fiber, food, boatbuilding, and basket making that was and is practiced by native peoples throughout California.

Vegetation crowds the banks of the pond as well. Part of it is shaded and protected by coyote bush and hazel, so small birds feel comfortable drinking and bathing in private. The other end is kept open for low-growing species, such as coast strawberry, sea bluff lettuce, sea pink, coast wallflower, and California aster *(Lessingia filaginifolia),* so it is full of flower, even in the fifth season. At the coast, these precious still, sunny, hot days, when we know that the turn of the year is coming, are fully exploited by all, beachgoers and butterflies alike.

AGAIN, EVERYTHING'S HERE

During the fifth season, ceanothus seeds explode all over my shady sitting area. The leaves, seeds, and twigs of ceanothus contain saponin, which makes a soapy lather when wet and was used for "baby washing" by the Yurok. When this debris falls into the little fountain nearby, the water becomes foamy. Landing on the glass table at which we sit, it becomes potential soap. I add water and make a lather, using dried pond scum as a sponge. In this way, I can clean the table using only materials that are at hand. For this task, everything's here.

32 Life Is Maintenance

Think of our life in nature,—daily to be shown matter, to come in contact with it,—rocks, trees, wind on our cheeks! The solid earth! The actual world! The common sense! Contact! Contact!

Henry David Thoreau, "Kraadn," 1848

It's good to know that what is needed most in the world, more even than food or warmth, is eloquence; it's good to know that most of life is maintenance of leaf-roofed huts—(and our own connection to the spirits)—and that it's a disaster to have a metal roof or to be saved once and for all.

Robert Bly, foreword to *Secrets of the Talking Jaguar,* 1998

GETTING SOME ON YOU

I once heard an apocryphal story about Tom Wolfe's reportorial journey with Ken Kesey and friends on the famous Electric Kool-Aid Acid Trip bus. At one point, the bus became mired in a muddy ditch. Wolfe stood by the side of the road in his immaculate clothes, taking notes on the event, while the rest of the crew, getting muddier and muddier, shoveled furiously. Finally, Kesey yelled, "Come on, Tom! Get some on ya!"

I think of that recommendation in late winter, when the first Indian lettuce seed is ready to harvest, and again in the spring, when the major thrust of weed removal takes place. I think of it in the late summer, the fifth season, when I go into the pond to divide tule, at which time the sensualities of this work exuberantly manifest themselves.

Maintenance is our chance to get some on us. Yard work, if planned so that it is not overly onerous, can be one of the best parts of this kind of gardening. If I am away from my restoration activities for more than a few weeks, it is as though a part of me has gone to sleep. I may not be aware of this dormancy until I return to the activities, the seed collecting and sowing, the transplanting, the vigilant eye for weeds, the coppicing and pruning, the basic, visceral engagement with the materials. Then I notice the relief of being back in the flow of it. The feel of mugwort in my hand, my gliding clippers, my sharp pruning saw, the tug of toyon berries detaching, are sensations that keep me connected and vital.

Paul Shepard, ecological thinker, in his book *The Others: How Animals Made Us Human* hypothesizes that our early contacts with animals, observing, imitating, and hunting, "made us human." "The human mind came into existence tracking. The animal world provided models for the very idea of thought." Shepard says that we became what we are in close contact with animals. Plants—their roots, stems, leaves, buds, flowers, and seeds—had a share in it too. Physical engagement with the vegetable world, essential for survival through humankind's evolution, also formed us. Restoration gardening can take us back to this world, through stimulating, humanizing activities, to be savored when kept manageable, to be relished as results become evident.

RENEWAL PRUNING

As is well documented by M. Kat Anderson and others, the landscape encountered by Europeans was the result of solicitous and meticulous attentions by the indigenous peoples, who gathered hundreds of different plants and animals in ways that allowed for their continued harvest. When the restoration gardener cuts her deer grass, western sword fern, or elderberry to the ground, she participates in a long-standing tradition of care.

The word *sustainable* might seem to imply the lack of necessity for maintenance, yet landscapes can never be fixed in time. To be truly sustainable, they must shift and flow in ways that involve us, which include activities that lead to renewal. I live in the scrublands, where fire was once a major agent of renewal. In its absence, I occasionally cut to the ground coyote bush, lizard tail, mugwort, elderberry, thimbleberry, willow, creek dogwood, sword fern, Douglas iris, sugar bush, gumplant, bush lupine, and anything else that thrives with renewal pruning. Along most creeks, flooding was a regular occurrence, depositing mud and debris around plants, and ripping them up and moving them downstream, sometimes to root in a new place. Wetland species are usually amenable to division by the gardener, and soil deposition can bring new life to creekside shrubs like willow, hazel, elderberry, and twinberry.

For many plants, the fifth season is a good time for renewal pruning. In August and September the quiet clicking of pruning shears, loppers, and hedgers is heard throughout the dream neighborhoods, making way for new shoots and sprouts that will emerge with the fall rains. Plants that are cut to the ground or whose dead parts are removed during this dry time are less vulnerable to the fungi and other invaders that come with the rains. When moisture does arrive, old cuts will have healed over, and new growth will be vigorous and unimpeded.

While training a new gardener, I tell her "You are fire, you are flood." I tell her to cut the eight-foot coyote bush to the ground, because it is full of dead branches and needs renewal. She looks nervous at this drastic measure, and I reassure her, but only seeing it for herself will be effective. She needs to see the fresh green growth after a fire, emerging from the massive roots, rising to cover the blackened stems. She needs to see the willow, buried six feet by the debris coming down the creek, putting out its fuzzy catkins. She hasn't seen yet how shapely they will rise.

In March and April, we who work here are pulling, digging, cutting, scraping, ripping, tearing, pickaxing, hoeing, weed whipping, and scything. We employ machetes, shovels, picks, pruning shears, English trowels, Japanese trowels, pick axes, hoes, sickles, rakes, and hands, gloved and ungloved. We relocate loads of biomass, consisting of oxalis, weedy grasses, wild geranium, scarlet pimpernel, hardly any thistles, and not as much African veldt-grass, sheep's sorrel, cape ivy, French broom, or Scotch broom as there used to be. Gaining an advantage, one determines not to lose it. Sometimes, though, it happens. A wet year can bring an explosion of moisture-loving weeds, like fennel, like poison hemlock, like cape ivy.

Ideally, weeds are removed before they have gone to seed, before rhizomes are strengthened by benign spring weather, before they can crowd out, shade out, or otherwise discourage all the spring, summer, and fall delights that are to follow. Though it happens every year, later it will be hard to remember, or even imagine, this level of activity. I initially resist this effort, but once I give in, there is satisfaction. Repeatedly, what seems an irritant becomes a nourishment.

YARD WORK

Ann Young is a landscaper-gardener who has spent many years and many hours working with native plants in a variety of situations. Unlike some landscapers, she doesn't hire out the weeding, pruning, and other maintenance work, but performs these tasks herself, paying close attention to the details and basing her maintenance strategies on her observation of these particular plants. There is something satisfying about watching a skilled and experienced horticulturist demonstrate techniques that are specific to and have evolved from the native plants with which she works.

Once bloom is passed and seeds have dropped, Douglas iris *(Iris douglasiana)* is a candidate for pruning attention. Its leaves frequently turn brown and may to some look unsightly. New leaves push up more easily when no dead leaves are in the way. Ann has often shown our workshop participants the way she handles Douglas iris when it is located in a prominent part of the garden where neatness is important.

Using double-bladed clippers, Ann begins cutting at the tip of the blade, and snips her way quickly down to the base, leaving about two inches of blade remaining. The small cut pieces of leaf blade are left where they fall, creating around the plant an attractive mulch of its own leaves. New shoots are free to push their way up through the ground when the fall rains come. Using this technique with other plants as well, Ann avoids moving butterfly and other insect larvae away from the nursery plants upon which they were laid and whose presence they may require for food and shelter.

In this way, she also avoids much of the labor-intensive movement of debris around the garden, carting and hauling being one of the main ways that gardeners spend energy. There is no need to move this debris to create a compost pile elsewhere, which is then carted back and spread around the plants from which the material came. When the material is too large and woody to be chopped up with clippers, Ann frequently uses long-handled loppers. In some situations, Douglas iris may be allowed to retain its cinnamon brown leaves through the fall and winter for a seasonal show. The combination of an active mind and repetitive activity can lead to insight into the particular ways that native plants "like" to be maintained and also, to flawless removal of invasive species.

Weeds are intimately known by Ann. I have watched her working through a deep woodland tangle of cape ivy, fingers sensitively searching through the duff for every purple root, ears listening for the right sound that means the entire rootlet was pulled free. What she doesn't want to hear is the "snap" that means a part of the root broke off, remains in the soil, and will be the parent of a new cape ivy plant. The softer sound of the entire root pulling free from the soil is the right kind of music to the ears.

SCALE OF ENDEAVOR

Design strategies, maintenance techniques, and reasonable goals can help keep garden work manageable, so that enjoyment rather than fatigue is the key emotion experienced. Scale of endeavor is an important component in this regard.

Alarm signals go up when I get a call from a customer who wants to seed fifteen acres with wildflower seeds. An army of weeders or a vat of herbicides spring to mind. Start small, I counsel. Most of us have limitations, whether financial, physical or time-related; a good rule of thumb is to take on significantly less than you think you can handle.

It sometimes happens that clients ignore my suggestions of moderation and my warnings against taking on too much, and stun me with what they accomplish. I watch with amazement when a certain kind of intense entrepreneurial, hyperactive, manic energy is directed towards restoration gardening. I've learned to sometimes get out of the way of others, whether human or western scrub-jay.

Still, for most of us, it can be helpful to state limited goals. "This much I will undertake this year, and this much only." A world of weeds surrounds us, and balance is the goal. I speak as one who has frequently taken on the unattended land that surrounds my garden. First try nudging, cajoling, educating, and pleading with neighbors to handle their own properties. Sometimes, the low-maintenance beauty of your own native garden will speak for itself.

LOW-MAINTENANCE GARDENS

Speaking up for the joyous necessity of physical interaction with plants, I eschew the desirability or possibility of so-called no-maintenance gardens. Low- rather than no-maintenance gardens are desirable to many, and can result from a number of choices. Plant size is one. A shrub or tree garden holds the territory with less attention from the gardener than a garden of small perennials. Since weeding holds prominence among gardening activities that require time, one strategy to lower maintenance is to plant shrubs and trees, which will shade out competitors with the exception of climbers like cape ivy or other weedy shrubs and trees, rather than smaller plants, which are vulnerable to many weeds. One option is to place plants with their ultimate height and width in mind, so that expensive and time-consuming pruning is not required. Alternatively, try "cramming" plants close together as a way to avoid weeding.

Tree gardens are low-maintenance ways to restore habitat. Trees, once established, hold territory with little work from us. Under the drip line of the tree, leaf litter can be its only ground cover, maintained with occasional weeding. Requiring no watering once established, no herbicides, no pesticides, and pruning only occasionally, correctly chosen trees make the simplest garden. Smaller understory trees, like vine maple *(Acer circinatum)* and California hazelnut

(Corylus cornuta var. *californica)*, can be planted at the drip line, though be aware that the drip line will be ever-changing as the tree grows.

Friends and acquaintances who may not think of themselves as gardeners, yet who still are responsible for some amount of land surrounding their houses, sometimes describe to me their gardening difficulties. They want advice, but may not like what I have to say. They may be seeking a way out of the requirements of the First Law of Gardening.

The law is this: The land requires our attention. Either you pay attention, or you hire somebody to pay attention, but attention, one way or the other, must be paid. "No-maintenance" gardens, if they include plants, generally use plants that in their ability to hold the space without human intervention are so vigorous as to be dangerous. Some of our worst invasive species, such as French broom, kudzu, Bermuda grass, kikuyu grass, and cape ivy, were horticulturally introduced in attempts to have no-maintenance gardens or lawns or to revegetate areas that had been damaged by mining. If no plants are used, even cement or asphalt will eventually fissure and break, weeds somehow finding their way to flourish in tiny cracks. Our maintenance activities can resemble in many cases the practices of the indigenous peoples who cared for this land before us. In our coppicing and pruning, our replanting, transplanting, and seeding, and in some situations, our burning, and also in our harvesting, we can be guided by those whose experience with this land preceded ours. Other major gardening guides can be the naturalists, who tell us when to stay our hand and let that dead limb remain so that it can be colonized, tunneled through, made into a cavity, or left to decay, who counsel us as to the uses of leaf litter covering the ground, and also, the need for bare soil.

TWISTING AROUND IN OUR SEATS

In spring 2004 the Society for Ethnobiology held its annual meeting in California. Joining the ethnobiologists as presenters were indigenous peoples from many California tribes, including basket weavers, land managers, flute makers, filmmakers, teachers, researchers, and elders. Certain messages were insistently repeated: the land that Europeans encountered was not an unmanaged wilderness, basketry and food plants require tending, diversity and land health declined when traditional land management technologies, particularly burning, were outlawed, and native knowledge and Western science can blend, each enriching the other.

At one talk, a traditional Karuk basket weaver discussed using fire to burn the species that produce the materials used in making basket caps, acorn storage baskets, gambling trays, winnowing trays, burden baskets, and cradle boards. Next to her at the presenters' table sat her young cousin, a research scientist whose dissertation analyzed the anatomical changes in basketry plants after burning. When it was her turn to present, her slides demonstrated microscopically what her cousin knew on the macro level, the way the new growth after burning forms tissue with the necessary strength and flexibility for good basketry materials. In these two family members were encapsulated some of the goals of the conference, a blending of native ways and Western science to help us bring back the health of the land and enrich our interactions with it.

At this conference, panels of presenters sat behind a table facing the audience, in back of them the screen upon which images were projected. During another talk, a botanist showed slides of soaproot *(Chlorogalum pomeridianum),* its roots and shoots emerging from seeds, the contractile roots pulling the bulb down into the soil. The other members of the panel, their backs to the screen, twisted around in their seats, the better to see what the botanist had to show them about this plant they knew through thousands of years of cultural intimacy. When basket weavers, soaproot brush makers, and flute makers showed images of the results of their work and detailed knowledge of the plants, professional botanists twisted around in theirs.

Sitting in the audience, I registered this spinal movement, one favored by yogis, reputed to foster vigor and health. It is one I often practice while sitting at my desk, a way to change my view while still holding onto my center. In this context, I took it as indicating an openness to and curiosity about the information and perspectives of all who were there to share knowledge of the plants and how to care for them better. It seemed a powerful moment in time, composed partly of the demonstrated willingness to twist around in our seats for a different view. For a brief moment during these three days of talks, the historical devastation of both land and people, the largely unheard stories of enslavement and genocide, the dismissal of the insights and values of complex, well-adapted cultures, the ensuing degradation of the land, and the concomitant bitter loss and regret, took a back seat. I had a vision of true allies recognizing each other at last, and our whole long, lovely state again attended to and cared for up and down its length, by those who know it through the plants.

To begin acquaintance with these plants by surrounding your home with them is the first step in a journey of great profundity and promise. The second is to care for them. This is the sublime nature of maintenance.

Coda We Are Not the First

> The faith being that if you make a commitment, and hang on until death, there are rewards.
>
> Wendell Berry, 2004

OKA YOUNG MAN'S RESTORATION PROJECT

A Yurok story of long ago, called *Oka Young Man's Deer Brush,* was told to A. L. Kroeber in 1902 by a blind Yurok storyteller named Tskerkr of Espeu. Kroeber describes Tskerkr as one of the best and most engaged narrators he encountered. "His narration was direct and incisive; that his feelings were powerful was evinced by the uncontrollable sobs into which the associations of his stories repeatedly shook him, when his massive frame would struggle and heave for minutes. . . . Sorrow, regret, longing, homesickness suffuse almost all of Tskerkr's tales." Perhaps not surprising given that, according to Kroeber, Tskerkr lived with his family in an otherwise deserted village, frequenting his sweat house entirely alone.

Tskerkr's tale bears a strong resemblance to modern restoration activities and motivations, intriguing us with old correspondences. The protagonist, Oka Young Man, purposefully fosters the deer, for hunting, and the birds, for the pleasure of their song, through restoring appropriate habitat. The story starts with a hunt and may have been a formula told to bring luck in hunting. It begins: "At Oka a young man always went to hunt. He saw no deer. But he continued hunting. There was no grass there, nor any open prairie. He thought, 'Perhaps the reason I see no deer is that there is no grass here. They do not like it as a place to live because there is nothing to eat.'"

Frustrated in the hunt, he goes far upstream in search of deer. "When he arrived, he saw a bush. The deer had been eating its leaves. . . . The bush had seeds. He wanted to take the seeds home to plant."

After collecting seeds and planting them, Oka Young Man enjoys a successful restoration project according to his projected goal of attracting deer. He is pleased with it ("Then he thought it was good"), but he doesn't stop there.

"So he went to where he had got the seeds, because he was always thinking of the birds, always wishing to hear the birds when he awoke, yet never hearing them. So he traveled upstream. When he arrived, that was the only place where he saw birds; he had never seen them where he himself grew. That is where he found the birds and the seeds, the only place at which he saw birds. Then he brought them downstream. In the morning he listened. Then he heard the birds outdoors. He thought, 'That is how I like it. I am glad to hear them.'"

Oka Young Man, now a thoroughly committed restorationist, and one with mythical powers we can only envy, goes to get more seed, and on the way sees the success of his earlier plantings. "Then when he went about, he saw deer tracks; yet he had never seen deer tracks before." He continues upstream to get water to make a spring for the deer and the birds, which they enjoy. "He heard all the birds shout; therefore he was satisfied. He thought, 'That is how I like it, to hear the birds shouting on all sides.' He felt well when he heard them."

When you hear the birds in your garden, may you too feel well.

A POSSIBLE BENT FOR THE MECHANISMS OF NATURE

As I peruse the literature of plants, animals, and people, a sensation of some poignancy may overtake me. It might happen when some piece of information from my own personal experience falls into place with a click, or when an indigenous observation or land management strategy reads like poetry. Or it may occur when some careful study from the world of Western science shines a light on some hitherto murky corner of the natural world.

Where the natural world is concerned, the news, as all must know or feel, is not good. In the face of a continual sense of emergency, it is the nature writers, past and present, and the biologists, naturalists, land managers, restorationists, and habitat gardeners who give me heart. The visionaries from the past, those who first stamped "Valuable. Handle with care" on this country's ecosystems, sometimes had premonitions of certain frightening aspects of the present state of things, as did the indigenous peoples, but they could not have realized how their words and actions would be models for us today. We have multiplied in their image, and we are many.

Aldo Leopold in 1939 wrote, "In a surprising number of men there burns a curiosity about machines and loving care in their construction, maintenance, and use. This bent for mechanisms, even though clothed in greasy overalls, is often the pure fire of intellect. It is the earmark of our times."

He went on to predict: "Everyone knows this, but what few realize is that an equal bent for the mechanisms of nature is a possible earmark of some future generation." It may be that we are that upcoming generation. As old stories like that of Oka Young Man tell us, we are not the first. Other old-time stories say that Scrub-Jay and Steller's Jay were given their acorn-planting work to do, while Eagle, Deer, Mountain Lion, and Human Beings each had their own assigned tasks.

Essayist Elaine Scarry says, "Beauty often comes to us through no work of our own, then leaves us prepared to undergo a giant labor." I send the energy that arises from these thoughts to my own work, as I ready a flat in which to sow the seeds of some small plant that is barely holding onto territory. I pour potting soil into the flat and smooth it out. Seeds are sown, and when seedlings emerge, they will be transplanted, watched, and aided when necessary. Hope is contained in this work, and hope pours through it. In my mind is a seasonal plan. In my mind is a plan.

ACKNOWLEDGMENTS

I wish to thank my sister Marjorie Larner, whose help on this book has been important, her presence in my life essential. The inspiration of her dauntless work in her chosen field of education goes deep, as we continue to discover the parallels between our arenas of endeavor. And with appreciation for her three sons, each in his own way larger than life: Cass, Joshua, and Alex Epstein. To my brother, Bernard Larner, with appreciation for his perspicacity, humor, and love, and to his amazing family: his wife, Leslee, and his children, Adam, Kate, Sean, and Sarah Rose, who was born a naturalist. To my esteemed cousins Paul Willen and Deborah Meier for the example of their stunning achievement, but equally for their quirky, honest, humorous takes on life and their dedication to making family reunions happen.

To Lillian Vallee, consummate writer, gifted poet, and innovative teacher, a Modesto Living Treasure, whose friendship is a gift. To M. Kat Anderson, both for her groundbreaking work, which combines scholarly research and years of fieldwork with great delicacy of insight and respect for the subject matter, and for the rich liveliness of our association, which I cherish. To Lester Rowntree the grandson for involving me in the work of Lester Rowntree the grandmother, in the process becoming family; also to Allison Green for taking me where otherwise I could not go. May more projects and trips present themselves. To Dave Fross, Bart O'Brien, and Carol Bornstein for their friendship and their landmark book *California Native Plants in the Garden.* To Arvind Kumar, of the Santa Clara Valley Chapter of the California Native Plant Society, whose dedication and perspicacity in his chosen avocation of native plant gardening are a particular encouragement to me. To the staff at the Point Reyes Bird Observatory, now known as PRBO Conservation Science, once of Stinson Beach, now of Petaluma. You are sorely missed; we were lucky to have you. In particular, I appreciate the work and help of the following: Tom Gardali, for vetting my drafts, steering me in the right direction, and making clear to me that which was hazy; Geoff Geupel, for his inspiring bird walks replete with countless quotable bits of wisdom; and Rich Stallcup, for his timely and gracious help with scrub-jay slides. Also to entomologist Robbin Thorp, for his work and taxonomical aid.

To Meg Simonds, skilled seed gatherer, indispensable office manager, and dear friend, who keeps the Larner Seeds office running and on an even keel. To Jasmin Clower, wearer of many hats, Larner Seeds–related and otherwise, with appreciation for both your broad perspective and your friendship. To the close friends of my home and my heart, indispensable in my life: Judy, Phil, Briana, and Bobby Buchanan; Bobby, Louie, and Kale Likover; Lea Earnheart, and Sasha and Tobi Earnheart-Gold; Ann Young and Michael Gaspars; Robert Levitt, Richard Diaz, and Rosie; Mary Nisbet and Tim Sheils; Victor, Sonia, and Kristina Amoroso; Patricia Wrobel Dickens; Mimi Calpestri; Linda Bennett; Charlie Adams; Sadja Greenwood; Jeff Creque; Lois and Manuel Talkowsky; Ken Botto; Waz Thomas; Ashley Ratcliffe; Lois Bridges; Lucinda Johnson; Meleta Kardos; Allen and Kathy Penny; Steve Hadland; Steve Lowry; and Keith Patterson.

To the memory of these departed ones, gone too soon: Robert Gold, Martina Johnson-Kent, Lynn Murray, and Nancy McDonald.

To my valued team at University of California Press: to project manager Dore Brown for her unflagging persistence, honed professional skills, and ability to make it all seem like fun, even late at night. Best of all, she became a convert. To Jenny Wapner, who displays great talent in her new position; you are gifted. To copyeditor Edith Gladstone for her helpful nudges in the right direction, and to Matthew Winfield and Chalon Emmons for support and weighing-in at opportune moments. To the accomplished and perceptive designer Sandy Drooker, with whom much of import took place over lunch. And to my ever-encouraging and supportive agent Andree Abecassis.

To my daughter, Dr. Molly Patterson, well on her way to healing a lot of people. Thank you for the pleasure you give me, in your life choices, and in all the aspects of our association, too many and precious to recount. You shine so brightly in my life.

To my husband, Peter Smith, finder of the fogstone, with whom I first saw wildflowers. When we are too old to go out to the flower fields, may there be children to make us our flower crowns. Until then, let us go ourselves, as frequently as possible.

I wish to express deep gratitude to the native peoples of California, past and present, for what they have shared, against great odds, and to particularly note

Tskerkr of Espeu, Lucy Young, Villiana Hyde, Tom Smith, Maria Copa Frias, Lucy Thompson, the Yokuts people who took in Jeff Mayfield, Campbell, the Sherwood Pomo who saved Edith Van Allen Murphey from drowning and helped her in her widowhood, Ace Hoagland, who showed her how to live in the mountains, the band of Pomo who educated and nourished the young Effie Hulbert, also George Laird, who rescued and taught his wife, Carobeth Laird, and also Essie Parrish, Mabel McKay, Elsie Allen, and Milton "Bun" Lucas who shared generously with so many. And to countless others, including many who are nameless but whose words and lives are with us still. In the present day, gratitude for the work of Ivan Jackson, Loren Smith, Ben Summerfield-Cunningham, Brian Bibby, Greg Sarris, Gaylen D. Lee, Julian Lang, Sage La Pena, Julia Parker, Lucy Parker, Linda Yamane, and many others, for their examples and words. May your work and lives go well.

NOTES

MY DREAM NEIGHBORHOODS

Epigraphs: William Dawson, *The Birds of California* (San Diego: South Moulton, 1923), 46; Lester Rowntree, "Wildflower Sanctuaries," *Desert Plant Life* 5, no. 11 (1934): 163; James Roof, "The Franciscan Region: Transcript of a Lecture," 1975, repr. in *Four Seasons* 2, no. 2 (October 2000): 49.

p. 5 *"In the last several decades"* Conservation International, "California Floristic Province," http://www.biodiversityhotspots.org/xp/Hotspots/california_floristic/conservation.xml.

p. 6 *a farmscape that weaves in the wild* Aldo Leopold, "The Farmer as a Conservationist," 1939; reprinted in Aldo Leopold, *For the Health of the Land: Previously Unpublished Essays and Other Writings,* ed. J. Baird Callicott and Eric T. Freyfogle (Washington, DC: Island Press/Shearwater Books, 1999), 161.

p. 7 *A friend of mine* Mary Bates Abbott, personal communication, 2002.

p. 9 *"They say, 'We made'"* James Roof, "The Franciscan Region."

p. 9 *In those days* Roof remarked that "oddly enough in San Francisco all the hilltops were crowned with coast Blue Blossom. And I wondered about that for a long time" (ibid., 47).

2 BIRDSONG RIPENS BERRIES

Epigraph: Paul Shepard, *The Others: How Animals Made Us Human* (Washington, D.C.: Island Press, 1996), 123.

p. 17 *In his last manuscript* Henry David Thoreau, *Wild Fruits: Thoreau's Last Discovered Manuscript,* edited and introduced by Bradley P. Dean (New York: W. W. Norton, 2000), 52.

p. 18 *"When people don't use the plants"* D. W. Peri and S. M. Patterson, "An Ethnographic and Ethnohistoric Survey and an Assessment of Native American Interests in the SUNEDCO Geothermal Leasehold, Mendocino County, California," report prepared for Ecoview Environmental Consultants, Napa, California, 1978 (on file at the

Ethnographic Laboratory, Sonoma State University). As quoted in Bev Ortiz, "Contemporary California Indian Basketweavers and the Environment," in *Before the Wilderness; Environmental Management by California Indians,* ed. Thomas C. Blackburn and M. Kat Anderson (Ballena Press, 1993), 199. See also Isabel Kelly, *Interviews with Tom Smith and Maria Copa: Isabel Kelly's Ethnographic Notes on the Coast Miwok Indians of Marin and Southern Sonoma Counties,* ed. Mary E. T. Collier and Sylvia Barker Thalman (San Rafael, CA: Miwok Archaeological Preserve of Marin, 1991).

p. 18 *"Today California Indians"* M. Kat Anderson, *Tending the Wild: Native American Knowledge and the Management of California's Natural Resources* (Berkeley: University of California Press, 2005).

p. 18 *Using one ripening to predict* Trevor C. Lantz and Nancy J. Turner, "Traditional Phenological Knowledge of Aboriginal Peoples in British Columbia," *Journal of Ethnobiology* 23, no. 2 (2003): 263–86.

p. 18 *As Paul Shepard says* Paul Shepard, *The Others: How Animals Made Us Human* (Washington, D.C.: Island Press, 1996), 123.

p. 20 *"The seeds [of grasses]"* Jennie Goodrich, Claudia Lawson, and Vana Parrish Lawson, *Kashaya Pomo Plants* (1980; Berkeley: Heyday Books, 1996), 85.

3 THE KEYNOTE BIRD

p. 22 *"the important functions of birds"* Kim Kreitinger and Tom Gardali, *Bringing the Birds Back: A Guide to Habitat Enhancement in Riparian and Oak Woodlands for the North Bay Region.* California Partners in Flight Regional Bird Conservation Plan no. 1, www.prbo.org/calpif. 2006, 8.

p. 23 The Birds of North America *The Birds of North America: Life Histories for the 21st Century.* This encyclopedic reference series of monographs jointly produced with the Cornell Lab of Ornithology and the Academy of Natural Sciences of Philadelphia describes, in detail, the life histories of all species of birds that breed in North America. The first account was produced in 1992, and vol. 18, with the last of the 716 accounts, appeared in 2002. With the completion of the final volume, the editorial office at the Academy of Natural Sciences was disbanded, but the project continues in a new, continually updated online format, at the Cornell Lab of Ornithology's Web site, http://bna.birds.cornell.eduBNA.

p. 23 *"the most privileged task"* and *"the very persons and lives"* William Leon Dawson, *The Birds of California* (San Diego: South Moulton, 1923), iii.

p. 24 *Among these volumes* Joseph Grinnell, "Uphill Planters," *Condor* 38 (1936): 80; and Emmett Hooper, "Another Jay Shoot in California," *Condor* 40 (1937): 162.

p. 24 *The bibliography from the monograph* Michael H. Macroberts and Barbara R. Macroberts, *Social Organization and Behavior of the Acorn Woodpecker in Central Coastal California* (Tampa, FL: American Ornithologists' Union, 1976).

p. 26 *prepared by the hands of native Californians* Bev Ortiz, *It Will Live Forever: Traditional Yosemite Indian Acorn Preparation* (Berkeley: Heyday Press in association with Rick Heide, 1991). According to Ortiz and Julia Parker, native Californians harvested acorns either by climbing into the tree to shake down nuts that were still held in this relatively sterile environment or by plucking them directly from the tree.

4 I LIVE IN A QUAIL YARD

p. 28 *"in the brush"* Isabel Kelly, *Interviews with Tom Smith and Maria Copa: Isabel Kelly's Ethnographic Notes on the Coast Miwok Indians of Marin and Southern Sonoma Counties,* ed. Mary E. T. Collier and Sylvia Barker Thalman (San Rafael, CA: Miwok Archaeological Preserve of Marin, 1991), 131.

p. 28 *"Their favorite and most easily acquired game birds"* Lowell John Bean, *Mukat's People: The Cahuilla Indians of Southern California* (Berkeley: University of California Press, 1972), 61.

p. 28 *"The brush country"* Jean François Galaup de la Pérouse, quoted by A. Starker Leopold, in *The California Quail* (Berkeley: University of California Press, 1977), 24.

p. 28 *"There I saw quail"* Walter Welch quoted in Leopold, *California Quail,* 26.

p. 28 *"wandered for an afternoon"* J. Smeaton Chase, *California Coast Trails: A Horseback Adventure from Mexico to Oregon* (1913; repr. Palo Alto, CA: Tioga Publishing, 1987), 244.

p. 29 *Today, with quail still a popular target* Jesse Garcia, personal communication, 2006. Of interest is the Department of Fish and Game's publication "Guide to Hunting Quail in California," by Sonke Mastrup, updated by D. Sam Blankenship and Jesse Garcia, July 2002, Department of Fish and Game, Wildlife Programs Branch, Upland Game Unit, Sacramento.

p. 30 *when Spaulding wrote about the need* Edward S. Spaulding, *The Quails* (New York: Macmillan, 1949).

p. 30 *"The basic component of the quail diet"* Leopold, *California Quail,* 51.

p. 33 *"When quail are flushed"* Ibid., 67.

p. 33 *"the well-loved book* That Quail, Robert" Margaret Stanger, *That Quail, Robert* (Philadelphia: Lippincott, 1966).

p. 34 *"Soon the whole flock"* Carobeth Laird, *Mirror and Pattern* (Banning, CA: Malki Museum Press, 1984), 5.

5 THE KEYNOTE PLANT

Epigraph: Lester Rowntree, *Flowering Shrubs of California and Their Value to the Gardener* (1939; repr. Stanford, CA: Stanford University Press, 1948), 285.

p. 38 *Also scrutinize the pages* Carol Bornstein, David Fross, and Bart O'Brien, *California Native Plants for the Garden* (Los Olivos: Cachuma Press, 2005), 21. With in-depth, useful, and eloquent descriptions of over five hundred plants and a carefully chosen picture for almost every entry, this landmark book is an invaluable tool.

p. 39 *Using another helpful book* John O. Sawyer and Todd Keeler-Wolf, *A Manual of California Vegetation* (Sacramento: California Native Plant Society, 1995).

p. 39 *they call for instant rounds of manzanita cider* Gaylen D. Lee, author of *Walking Where We Lived: Memoirs of a Mono Indian Family* (Norman: Oklahoma University Press, 1998), describes his family's method for making manzanita cider: Clean and dry the berries. Crush them with a heavy implement (I use a potato ricer), and then put them into a strainer lined with cheesecloth. Pour roughly twice the amount of water as of berries slowly through the cloth and strainer, catching the liquid in a glass jar.

Lee notes that some individuals within a population have berries that are sweeter than others.

6 SING WILLOW

Epigraph: Erwin G. Gudde, *California Place Names,* 4th ed. (Berkeley: University of California Press, 1998), 424.

p. 42 *"I've eaten parts of some"* William Least Heat-Moon, *PrairyErth: A Deep Map* (Boston: Houghton Mifflin, 1991), 238.

p. 42 *Employed to quiet the agitation of arthritis* Drug Digest, Drug Library—Drugs & Vitamins, www.drugdigest.org. Note that there is a category of persons called "salicylate-sensitive," and that liver and kidney problems, gastrointestinal problems, bleeding disorders, and being a child are all reasons not to take willow bark. Tannins are also present, which can be harmful over the long haul. If you still wish to try it, you may brew dried white willow bark *(Salix alba)* into tea by pouring 5 ounces of boiling water over 1,000 to 3,000 mg of shredded white willow bark and letting it soak for five minutes. Other recipes for medicinal use of this non-native species abound, though none specifiying the use of indigenous willows could be found.

p. 43 *"it is very exceptional"* Christopher Newsholme, *Willows: The Genus Salix* (London: B. T. Batsford, 1992), 34.

p. 43 *Birders go to willows to find* J. Grinnell and M. W. Wythe, "Bird Life of the San Francisco Bay Region," *Pacific Coast Avifauna* (Cooper Ornithological Club, Berkeley, California, 1927): 18. These authors found the yellow-breasted chat to be fairly common to the Bay Area in the twenties, nesting in willow thickets "bordering the lower, sluggish courses of streams." In southern California, this bird was found in willows in early seral stage along streams and around ponds.

p. 44 *To deepen your understanding of willows* George Wharton James, *Indian Basketry,* 4th ed. (1909; repr. New York: Dover Publications, 1972).

pp. 44 and 45 *Some species are more useful* and *Native Californians also used willow* Beatrice M. Beck and Sandra S. Strike, *Ethnobotany of the California Indians,* vol. 2, *Aboriginal Uses of California's Indigenous Plants* (Champaign, IL: Koeltz Scientific Books, 1994).

p. 47 *"Windbreaks treated in this way"* Newsholme, *Willows,* 40.

7 THE LANDSCAPING IDEAS OF JAYS

Epigraphs: Joseph Grinnell, "Uphill Planters," *Condor* 38 (1936): 81; William R. Jordan III, *The Sunflower Forest* (Berkeley: University of California Press, 2003), 24.

p. 49 *"In that place and on that day"* Grinnell, "Uphill Planters," 80.

p. 50 *"Doubtless . . . this miserly trick"* William Leon Dawson, *The Birds of California* (San Diego: South Moulton, 1923), 50.

p. 50 *"The Middle Mewuk"* *The Dawn of the World: Myths and Tales of the Miwok Indians of California,* collected and edited by C. Hart Merriam (1910; Lincoln: University of Nebraska Press, 1993), 213.

p. 50 *Of the nine coast live oaks now growing here* "Western Scrub-Jays are purported to be important agents for dispersal of several oak and pine species, but extent of tree recruitment resulting from jay nut storage has never been measured." Robert L. Curry, A. Townsend Peterson, and Tom A. Langen, "The Western Scrub Jay," in *The Birds of North America: Life Histories for the 21st Century,* Cornell Laboratory of Ornithology and Academy of Natural Sciences monograph no. 712, 2002.

p. 51 *Clark's nutcracker* (Nucifraga columbiana) Ronald Lanner, *Made for Each Other: A Symbiosis of Birds and Pines* (New York: Oxford University Press, 1996).

p. 51 *"Working in concert"* Lanner, *Made for Each Other,* 3.

p. 51 *"He didn't do half as well"* Russell P. Balda, quoted in Candace Savage, *Bird Brains: The Intelligence of Crows, Ravens, Magpies, and Jays* (San Francisco: Sierra Club Books, 1995), 119.

p. 51 *"facilitate activity by standing still"* Curry et al., "Western Scrub Jay," in *Birds of North America.*

p. 52 *Experiments have shown* Peter A. Bednekoff, Russell P. Balda, Alan C. Kamil, and Arla Hile, "Long-term spatial memory in four seed-caching corvid species," *Animal Behavior* (1997): 53, 335–41.

p. 52 *Innocent scrub-jays* N. J. Emery and N. S. Clayton, "Effects of Experience and Social Context on Prospective Caching Strategies by Scrub Jays," letters to *Nature, Nature* 414 (November 2001): 443.

p. 52 *"He, the malaprop"* Dawson, *Birds of California,* 47. Still speaking of the California jay, Dawson calls him "the sweet, authentic devil."

p. 52 *"undoubtedly the chief biological control factor"* Ibid., 47.

p. 52 *"This is likely not so"* Tom Gardali, personal communication, 2005.

p. 52 *"Use the money you would otherwise spend"* Geoffrey Geupel, personal communication, 2005.

p. 53 *"Jays and other corvids"* Gardali, personal communication, 2005.

p. 53 *"The whole effort"* Harold and Josephine Michener, "California Jays, Their Storage and Recovery of Food, and Observations at One Nest," *Condor* 47 (1945): 206.

p. 53 *"Everything seemed peaceful"* Ibid.

p. 53 *Jay hunts organized in the 1930s* Emmett Hooper, "Another Jay Shoot in California," *Condor* 40 (1937): 162.

8 IN PRAISE OF THE UNLEAVING

Epigraphs: Gerard Manley Hopkins, "Spring and Fall; To a Young Child," found on www.poets.org; John Burroughs, 1913 poetry found on www.brainyquote.com/quotes/authors/j/john_burroughs.html.

p. 57 *Wrentits no longer breed* Personal communication, 2004, from Geoffrey Geupel, author of "The Wrentit," in *The Birds of North America: Life Histories for the 21st Century,* Cornell Laboratory of Ornithology and Academy of Natural Sciences monograph no. 654, 2002.

p. 58 *One friend . . . welcomes . . . "the three poisons"* Marion Weber, personal communication, 2003.

p. 61 *"As an added benefit"* Ray Raphael, *Tree Talk* (Covelo, CA: Island Press, 1981), 89. This book and its revised, expanded edition (*More Tree Talk* [1994]) are an essential resource for the exurban or rural landowner in northern California. Through interviews with people intimately associated in many capacities with trees and land, Raphael presents an illuminating picture of the problems and possibilities past and present of timber harvest.

9 EATING THE RAIN

Epigraph: Leslie L. Haskin, *Wild Flowers of the Pacific Coast* (Portland: Binfords and Mort, 1934), 89.

p. 65 *During the gold rush* Kat Anderson, personal communication, 2005.

p. 66 *The recent upsurge of interest* Some of the current popular research on phytochemicals can be found in these two books: James A. Joseph, Daniel A. Nadeau, and Anne Underwood, *The Color Code: A Revolutionary Eating Plan for Optimum Health* (New York: Hyperion Press, 2002); and Steven Pratt and Kathy Matthews, *Superfoods: Fourteen Foods That Will Change Your Life* (New York: HarperCollins, 2004). Also, for a more philosophical look at plant constituents and what they offer humankind, I recommend Steven Buhner, *The Lost Language of Plants: The Ecological Importance of Plant Medicines to Life on Earth* (White River Junction, VT: Chelsea Green, 2002).

p. 68 *"Did I choose"* Michael Pollan, *The Botany of Desire: A Plant's Eye View of the World* (New York: Random House, 2001).

p. 69 *"Pedicels reflexed"* *The Jepson Manual: Higher Plants of California,* ed. James C. Hickman (Berkeley: University of California Press, 1993), 898.

p. 70 *"Of the wild lettuce"* Stephen Powers, *Tribes of California,* introduction and notes, Robert F. Heizer (1877; Berkeley: University of California Press, 1976), 7.

10 ETERNAL VIGILANCE

Epigraph: Jake Sigg, "Oxalis pes-caprae," Species Focus, *Cal EPPC News* (now *Cal IPC News*) 11, no. 1 (Spring 2003).

p. 72 *Though previously divided* For a story on "green cemeteries," where land preservation may be part of the package, see Tad Friend's article in the *New Yorker,*

August 29, 2005, "The Shroud of Marin," about current attempts to cater to those with environmental concerns about their final resting places.

p. 72 *I am told* The presence of significant accumulations of the stored parts of Bermuda buttercup in gopher runs has been reported by both Mike Koslowski and Jeffrey Caldwell. Respective personal communications, Koslowski to Jake Sigg, January 2005, and Caldwell to Judith Lowry, 2004.

p. 73 *"Oxalis gives a new meaning"* Jake Sigg, personal communication, March 2005.

11 RAIN-TIME READING

p. 74 *The Nez Perce tribe of Idaho told a child* Nu Mee Poom Tit Wah Tit, *Nez Perce Legends* (1972), as quoted in Joseph Bruchac, *Roots of Survival: Native American Storytelling and the Sacred* (Golden, CO: Fulcrum Publishing, 1996).

p. 74 *"If stories should be retold"* William Guy Spittall, introduction to *Myths of the Iroquois,* by E. A. Smith, U.S. Bureau of American Ethnology, 2nd annual report for 1880–81 (Washington DC, 1883), as quoted in Bruchac, *Roots of Survival,* 70.

p. 75 *Abenaki Indian Joseph Bruchac* Bruchac, *Roots of Survival,* 77.

p. 75 *Greg Sarris, Coast Miwok* Greg Sarris, *Keeping Slug Woman Alive: A Holistic Approach to American Indian Texts* (Berkeley: University of California Press, 1993), 22, 194.

p. 75 *"My grandparents said"* Gaylen D. Lee, *Walking Where We Lived: Memoirs of a Mono Indian Family* (Norman, OK: University of Oklahoma Press, 1998), 12.

p. 76 *"My mother belonged"* Lester B. Rowntree (Rowntree's grandson), personal communication, 2006.

12 LESTER ROWNTREE

Epigraph: James Roof interview with Rosemary Levenson in *Lester Rowntree: California Native Plant Woman* (Regional Oral History Office, Bancroft Library, University of California, Berkeley, 1979), 79.

p. 78 *We know Rowntree in the main* *Hardy Californians* (1936; new, expanded ed., Berkeley: University of California Press, 2006); and *Flowering Shrubs of California and Their Value to the Gardener* (1939; repr. Stanford, CA: Stanford University Press, 1948).

p. 79 *"I drool with nostalgia"* Rowan Rowntree interview in Levenson, *Lester Rowntree,* 315.

p. 81 *"But before the season's climax is reached"* Rowntree, *Hardy Californians,* 2.

p. 82 *"To the conversant"* Levenson, *Lester Rowntree,* 75.

p. 82 *"Losing all that work"* Roof interview in ibid., 76.

p. 83 *American horticulture would take longer strides"* Rowntree, *Hardy Californians,* lxxx (page number refers to the 2006 edition).

p. 84 *"Particularly during the past five years"* Rowntree, *Flowering Shrubs,* 282.

p. 85 *"There is good chaparral"* Ibid., 20.

p. 86 *"I have put down what I have gleaned"* Ibid., v.

p. 86 *"Grow* Myrica californica*"* Ibid., 267.

p. 87 *"Modern boys and girls"* Lester Rowntree, "Lone Hunter," *Atlantic Monthly* 163 (1939): 809–16.

p. 87 *James Roof describes her frustration* Roof interview in Levenson, *Lester Rowntree,* 50.

p. 88 *In 1944* J. S. Dixon, "California Jay Picks Ticks from Mule Deer," *Condor* 52 (1944): 140–41.

p. 89 *"One of the best* [Senecio]*"* Rowntree, *Hardy Californians,* 32.

p. 90 *"One of the greatest advantages"* Sydney Mitchell, *Your California Garden and Mine* (New York: M. Barrows, 1947), 13.

p. 90 *"wild floods of blossoms"* Lester Rowntree, "On a Trail of Beauty," in *The Sierra Nevada, the Range of Light,* ed. Roderick Peattie (New York: Vanguard Press, 1947), 95.

p. 91 *In 2006, they undertook* Lester Rowntree, *Hardy Californians: A Woman's Life with Native Plants,* new, expanded edition, ed. Lester B. Rowntree (Berkeley: University of California Press, 2006).

13 EDITH VAN ALLEN MURPHEY

Epigraph: Skee Hamann, biography, "Edith Van Allen Murphey," written between 1980–83 and archived at the Held-Poage Memorial Research Library in Ukiah, California.

p. 92 *I first encountered Edith's book* Edith Van Allen Murphey, *Indian Uses of Native Plants* (Ukiah: Mendocino County Historical Society, 1959), n.p.

p. 92 *Shortly thereafter, I found* Skee Hamann, "Seed Seeker of the Flowering West," *Horticulture* magazine, March 1978.

p. 92 *"A Day among Californian Wildflowers"* Lester Rowntree, "A Day among Californian Wildflowers," Better Homes and Gardens Leaflet no. B-G-53, date unknown.

p. 93 *For two years"* Hamann, "Seed Seeker," 15.

p. 94 *"I especially was told"* Ibid., 12.

p. 94 *Her tomboy nature encouraged by her father* Hamann, biography. The vigor and detail of this first manuscript formed my initial encounter with Murphey. I have taken most of my biographical information from it, and Hamann's unpublished biography of Murphey is the source of parenthetical page citations given in chapter 13. Hamann was encouraged to rewrite the manuscript for a potential New York publisher, which manuscript is also available at Held-Poage Memorial Library. A book now exists based on this later manuscript, titled *Bahai Wakidu, the Seed Seeker: A Biography of Edith Van Allen Murphey,* ed. Rebecca Snetselaar and Marcia Platt, Mendocino County Museum Grassroots History Publications no. 21 (Willits, CA, 2004). This book, the culmination of years of research by the editors, rounds out Edith's story with wonderful photographs and invaluable footnotes.

p. 98 *"The hurt of her grief"* Skee Hamann had need of such methods of handling tragedy. Her own daughter was brutally murdered, and Hamann herself fell ill before many of her projects could come to fruition (she died of cancer). Sensitive to the accomplishments and unusual personalities of her subjects, this generous and insightful biographer of botanists has left an important legacy herself. Aside from "Seed Seeker," her articles include "Edith Murphey, Pioneer Botanist in Mendocino," *Fremontia* (July 1984): 15–18; "Two Pioneer Botanizers [Ruby and Van Deventer] in Del Norte County," *Fremontia* (January 1982): 3–16; "The Wildflower Lover [Lester Rowntree] at Ninety-seven," *Fremontia* (January 1976): 3–16. Her subjects made Hamann well aware of how productive the later years might be. As Edith put it in her statement for Edward R. Murrow's *This I Believe,* "Truly all the sugar is in the bottom of the cup."

p. 100 *These painful stories* "Out of the Past: A True Indian Story Told by Lucy Young of Round Valley Indian Reservation, to Edith V. A. Murphey," *California Historical Society Quarterly* 20, no. 4 (December 1941): 380.

p. 102 *"botanizing joyfully around the West"* Attributed to Ben Hur Lampman, editor of the *Portland Oregonian* and quoted in *Bahai Wakidu,* 83n.

14 GERDA ISENBERG

Epigraph: The source of the quote on Gerda Isenberg's bookplate is John Burroughs, "The Summit of the Years," 1913.

p. 103 *"Yerba Buena Nursery has long been"* Bart O'Brien quoted in *Yerba Buena Memories* (1997).

p. 104 *"They always focus on my work boots"* Gerda Isenberg, personal communication, 1979.

p. 105 *"Gerda is someone"* Suzanne B. Riess, who conducted the interviews for *Gerda Isenberg: California Native Plants Nurserywoman, Civil Rights Activist, and Humanitarian,* with an introduction by Bart O'Brien (Regional Oral History Office, Bancroft Library, University of California, Berkeley, 1991).

p. 106 *"My relatives wore me down"* Gerda, personal communication, 1993.

p. 107 *"I was always trying"* Ibid.

p. 107 *"five adults and six children"* Ibid.

p. 109 *I would rush into the library* Forest Service, U.S. Department of Agriculture, *The Woody Plant Seed Manual* (Caldwell, NJ: Blackburn Press, 2002).

p. 113 *On June 11, 1996* *Yerba Buena Memories.*

15 FOREST GARDENS

Epigraph: Chris Maser, *The Redesigned Forest* (San Pedro, CA: R. and E. Miles, 1988).

p. 115 *As some of us move* Ray Raphael, *More Tree Talk: The People, Politics, and Economics of Timber* (Washington, D.C.: Island Press, 1994).

p. 120 *Total contact with the forest floor* On the five decay classes and on coarse woody debris in general, see Chris Maser and James Trappe, eds., *The Seen and Unseen World of the Fallen Tree,* General Technical Report PNW-164 (Portland: Pacific Northwest Forest and Range Experiment Station, U.S. Dept. of Agriculture, Forest Service, 1984), 6. Also see Chris Maser, Steven P. Cline, Kermit Cromack Jr., James M. Trappe, and Everett Hansen, "What We Know about Large Trees That Fall to the Forest Floor," in Chris Maser, Robert F. Tarrant, James M. Trappe, and Jerry Franklin, eds., *From the Forest to the Sea: A Story of Fallen Trees,* General Technical Report PNW-GTR-229 (Portland: Pacific Northwest Forest and Range Experiment Station, U.S. Dept. of Agriculture, Forest Service, 1988), 32; table 2.6 describes a five-class system of decay based on fallen Douglas fir trees.

p. 121 *"Habitats provided by the death"* Maser et al., "What We Know about Large Trees That Fall to the Forest Floor," in *From the Forest to the Sea,* 44.

p. 121 *The real consequences* According to Chris Maser et al., "About 140 years are needed for nutrients to cycle in large, fallen trees and more than 400 years for such trees to become incorporated into the forest floor; they therefore interact with the plants and animals of the forest floor over a long period of forest and stand successional history." Maser et al., ibid., 37.

p. 121 *Some of the early works* Maser et al., *From the Forest to the Sea;* Maser and Trappe, eds., *The Seen and Unseen World of the Fallen Tree;* Maser, *The Redesigned Forest;* and Chris Maser, *Forest Primeval: The Natural History of an Ancient Forest* (San Francisco: Sierra Club Books, 1989). See also Jerry Franklin and Chris Maser, *Ecological Characteristics of Old-Growth Douglas-Fir Forests* (Portland: Pacific Northwest Forest and Range Experiment Station, U.S. Dept. of Agriculture, Forest Service, 1981). For an excellent overview, see David Kelly, *Secrets of the Old Growth Forest* (Salt Lake City: Peregrine Smith Books, 1988), illustrated by Gary Braasch.

p. 123 *This element* See *The Salmon Forest,* DVD, written, produced, and directed by Caroline Underwood (CBS Enterprises, 2001; available through Bullfrog Films, Oley, PA, 2002, www.bullfrogfilms.com).

16 THE FLOWER DANCE

Epigraphs: Lester Rowntree, "Wildflower Sanctuaries," *Desert Plant Life* 5, no. 11 (1934): 163; Wallace Hebberd, *The Wild Gardens of Old California* (Santa Barbara: CFS, 1927), 23.

p. 128 *"In the blossom time"* William Shipley, ed. and trans., *The Maidu Indian Myths and Stories of Hanc'Ibyjim* (1927; repr. Berkeley: Heyday Books, 1991), 12.

p. 129 *"The magnificent sea of wild-flower blooms"* Anna H. Gayton, *Yokuts and Western Mono Ethnography,* Anthropological Records, vol. 10, no. 1 (Berkeley: University of California, 1948); facsimile reprint available through Coyote Press, in Salinas, CA.

p. 132 *"Some of these seeds are very beautiful"* David Prescott Barrows, *Ethno-Botany of the Coahuilla Indians of Southern California* (1900; repr. Banning, CA: Malki Museum Press, 1977).

p. 132 *"No, but my grandmother did"* Renee Shahrokh, "The Traditional Processing of Red Maids, a Native Wild Seed: Harvesting, Winnowing and Roasting with Hot Coals in Baskets" (talk presented to the annual meeting of the Society for Ethnobiology, Davis, CA, spring 2004). As a teacher and researcher at Sacramento State University, Shahrokh is experimenting extensively with the technology of red maid seed collection and preparation.

p. 133 *"The weeds introduced since the coming of white people"* Pliny Earle Goddard, *Life and Culture of the Hupa,* University of California Publications in American Archaeology and Ethnology (Berkeley: University of California Press, 1903).

p. 133 *Or perhaps it too only awaits revival* For ongoing cultural activities among California's native peoples, see Heyday Books' *News from Native California,* Berkeley, 1987–2005.

p. 134 *"As we passed below the hills"* Thomas Jefferson Mayfield, *Tailholt Tales,* by Frank F. Latta (Santa Cruz, CA: Bear State Books, 1976), 13.

p. 134 *"Wiping away more tears"* Steve Wall with Harvey Arden, *Wisdom's Daughters: Conversations with Women Elders of Native America* (New York: HarperCollins, 1994).

p. 134 *"This and future generations"* Nicolus Hanson, *As I Remember* (Willows, CA, 1944).

p. 138 *"From the old camping place"* In Herbert Luthin, *Surviving Through the Days: Translations of Native California Stories and Songs* (Berkeley: University of California Press, 2002).

p. 138 *"The Flower Lover"* Chris Simon, *Down an Old Road: The Poetic Life of Wilma Elizabeth McDaniel,* videocasette (2001; available from Chris Simon, *csimon@lasal.net*). McDaniel, a living treasure, has been a prolific poet, and her poems have been collected in a number of volumes by a number of poetry presses. Some of her books include *A Primer for Buford* and *Borrowed Coats* (Brooklyn: Hanging Loose Press, 1989 and 2001, respectively); *Hoeing Cotton in High Heels* (Austin: Liquid Paper Press, 2002); and *Getting Love Down Right* (Springville, CA: Back 40 Publishing, 2000).

p. 139 *Call the Wildflower Hotlines* Anza Borrego Desert State Park, 760-767-4684, www.anzaborrego.statepark.org; Antelope Valley California Poppy Reserve, 661-724-1180, www.calparksmojave.com; Death Valley National Park, 760-786-2331, www.death.valley.national-park.com; Joshua Tree National Park, 760-367-5500, www.joshua.tree.national-park.com; Theodore Payne Foundation, 818-768-3533, www.theodorepayne.org.

17 THE WEED DANCE

Epigraph: Sarah Connick and Mike Gerel, "Don't Sell a Pest," *CAL-IPC News* 13, no. 2 (Summer 2005).

p. 141 *The California Invasive Plant Council* Connick and Gerel, "Don't Sell a Pest."

p. 142 *Acknowledging her right* Carla C. Bossard, John M. Randall, and Marc C. Hoshovsky, eds, *Invasive Plants of California's Wildlands* (Berkeley: University of California Press, 2000).

p. 144 *The botanist John Klironomos* J. N. Klironomos, "Feedback with Soil Biota Contributes to Plant Rarity and Invasiveness in Communities, "*Nature* 417 (May 2002):

67–70; and Wim H. Van Der Putten, "Plant Population Biology: How to Be Invasive," *Nature* 417 (May 2002): 3–33.

p. 145 *"It is action, the language of nature"* William R. Jordan III, *The Sunflower Forest* (Berkeley: University of California Press, 2003), 92.

18 DO YOU TALK TO PLANTS?

Epigraph: Paul Shepard, *The Others: How Animals Made Us Human* (Washington, D.C.: Island Press, 1996), 8.

p. 148 *"Please note that my pamphlet"* Craig Dremann, *The Redwood Forest and the Native Grasses and Their Stories, as Told to Craig C. Dremann* (Redwood City: Redwood City Seed, 1987).

p. 148 *Milicent Lee recorded a story* Milicent Lee, *Indians of the Oaks* (1937; repr. Ramona, CA: Acoma Books, 1978), 102.

p. 149 *Tom Brown, wilderness survival teacher* Tom Brown, *Tom Brown's Guide to Wild Edible and Medicinal Plants* (New York: Penguin, 1986).

p. 149 *Joseph Bruchac, an Abenaki Indian* Joseph Bruchac, *Roots of Survival: Native American Storytelling and the Sacred* (Golden, CO: Fulcrum Publishing, 1996).

p. 149 *Patrick Orozco, a Rumsien Ohlone* Patrick Orozco, "An Ohlone Story," *Terrain Magazine* (Berkeley Ecology Center/San Bruno Mountain Watch Newsletter, Brisbane, CA), Spring 2003.

p. 150 *"sharing of soul essences"* Stephen Buhner, *The Lost Language of Plants: The Ecological Importance of Plant Medicines to Life on Earth* (White River Junction, VT: Chelsea Green Publishing, 2002).

19 ROCK KNOWS

p. 151 *Garden Conservancy's Heritage Garden Tour* The Garden Conservancy, PO Box 219, Cold Spring, NY 10516 (845) 265 2029; e-mail info@gardenconservancy.org. "The Garden Conservancy is a national, nonprofit organization founded in 1989 to preserve exceptional American gardens for the public's education and enjoyment."

p. 153 *Robert M. Thorson* Robert M. Thorson, *Stone by Stone: The Magnificent History in New England's Stone Walls* (New York: Walker, 2002).

p. 154 *The Going Native Garden Tour* For information on tours of native plant gardens in the South Bay, contact info@GoingNativeGardenTour.com. For information on the list that started it all, go to www.calypteanna.com/ca-natives.html.

p. 154 *Inspired by the South Bay experience* To learn about the annual free self-guided tour of Alameda and Contra Costa County gardens sponsored by Bringing Back the Natives, go to www.BringingBackTheNatives.net.

p. 155 *Native plant gardens in Los Angeles and Orange Counties* The Theodore Payne Foundation for Wildflowers and Native Plants hosts an annual garden tour in Los Angeles; infomation can be found at www.theodorepayne.org. Information on the Orange County Native Garden Tour can be obtained through the Orange County Chapter of the California Native Plant Society at www.occnps.org.

p. 155 *Kumar believes that native plant horticulture* Kumar's suggestions on how to organize a garden tour can be found in Arvind Kumar, "Organizing a Native Garden Tour," *Fremontia* 34 (January 2006): 17.

20 WHERE'S THE CLOVER?

Epigraphs: Carolina Welmas, *Muluwetam* (Banning, CA: Malki Museum Press, 1973), 48; Villiana Calac Hyde and Eric Elliott, *Yumáyk Yumáyk: Long Ago* (Berkeley: University of California Press, 1994), 549.

p. 158 *"In people, increases in cancer"* Stephen Buhner, *The Lost Language of Plants: The Ecological Importance of Plant Medicines to Life on Earth* (White River Junction, VT: Chelsea Green Publishing, 2002), 206.

p. 159 *clovers "were a mainstay"* M. Kat Anderson, *Tending the Wild: Native American Knowledge and the Management of California's Natural Resources* (Berkeley: University of California Press, 2005), 269.

p. 160 *"When pioneers move into a new area"* Joseph L. Chartkoff and Kerry Kona Chartkoff, *The Archaeology of California* (Stanford, CA: Stanford University Press, 1984), 41.

p. 161 *"California natives"* Beatrice M. Beck and Sandra S. Strike, *Ethnobotany of the California Indians,* vol. 2, *Aboriginal Uses of California's Indigenous Plants* (Champaign, IL: Koeltz Scientific Books, 1994), 132.

p. 161 *"grazing like animals"* Anderson, *Tending the Wild,* 269–70.

p. 161 *One example of combining food* Strike, *Ethnobotany of the California Indians,* 131.

p. 161 *"This passing of useful botanical knowledge"* William Least Heat-Moon, *PrairyErth: A Deep Map* (Boston: Houghton Mifflin, 1991), 239.

p. 162 *Kelly's interviews* *Interviews with Tom Smith and Maria Copa: Isabel Kelly's Ethnographic Notes on the Coast Miwok Indians of Marin and Southern Sonoma Counties,* ed. Mary E. T. Collier and Sylvia Barker Thalman (San Rafael, CA: Miwok Archaeological Preserve of Marin, 1991), 55, 58, 121.

p. 162 *In 1993 a single plant* "Close Call Takes Toll on Rare Clover," *Science News* 155 (January 16, 1999).

21 SAYING FAREWELL TO SPRING

Epigraph: Leslie L. Haskin, *Wild Flowers of the Pacific Coast* (Portland: Binfords and Mort, 1934).

p. 168 *Ringing countless changes on the color pink* *The Jepson Manual: Higher Plants of California,* ed. James C. Hickman (Berkeley: University of California Press, 1993), describes its considerable variety this way: "Petals 5–60mm, often lobed or clawed, lavender or pink to dark red, pale yellow, or white, often spotted, flecked, or streaked with red, purple or white" (786). Another taxonomist, John Thomas Howell, in *The Marin Flora,* says of *Clarkia amoena,* "On the hills immediately adjacent to the coast the flowers are unusually large and the color is a very attractive pink" (200). Rowntree called this annual "lovely *Clarkia pulchella* var. *alba*" and touted its performance in her New Jersey garden. Lester Rowntree, "White Clarkia," *Horticulture* 28, no. 3 (1950): 107. *Clarkia pulchella* is given the common name of "pink fairies" by Bertha M. Rice and Roland Rice, in *Popular Studies of California Wild Flowers* (San Francisco: Upton Bros. and Delzelle, 1920).

p. 168 *"large loaves of mashed"* M. Kat Anderson, *Tending the Wild: Native American Knowledge and the Management of California's Natural Resources* (Berkeley: University of California Press, 2005), 264. "Farewell-to-spring and redmaids were plentiful in the Clipper Gap and Auburn area before 1900, enabling the Southern Maidu to make ten-to-fifteen-pound loaves of mashed, cooked seeds."

p. 168 *"Mashed farewell-to-spring"* Brian Bibby, *Deeper than Gold: Indian Life in the Sierra Foothills* (Berkeley: Heyday Books, 2005), 122. According to David Roche, a Sierra Miwok, these personal names, "nearly always represent action, being based on verbs rather than nouns, and the true meaning of a name is implied rather than literal." Relevant here is the implied importance of the qualities of the food made from the seeds of *Clarkia* species.

p. 168 *The fragrance of this small flower* "On the Scent of Designer Blooms," (London) *Daily Telegraph,* July 10, 2002.

p. 169 *"lavender shading paler near middle"* Greg Wahlert and Phil Van Soelen, "The Conservation of Two Sonoma County Manzanitas," *Fremontia* 33, no. 3 (July 2005).

p. 169 *The Sonoma County Milo Baker Chapter* Eugene N. Kozloff and Linda H. Beidleman, *Plants of the San Francisco Bay Region: Mendocino to Monterey* (Pacific Grove, CA: Sagen Press, 1994). A new, revised edition has been published as Linda H. Beidleman and Eugene N. Kozloff, *Plants of the San Francisco Bay Region: Mendocino to Monterey* (Berkeley: University of California Press, 2004).

p. 171 *Other* Clarkia *species* Helen K. Sharsmith, *Spring Wildflowers of the San Francisco Bay Region* (Berkeley: University of California Press, 1965, repr.).

p. 171 *The need for botanic names* Kozloff and Beidleman, *Plants of the San Francisco Bay Region; Jepson Manual;* Earl J. Larrison, Grace W. Patrick, William H. Baker, and James A. Yaich, *Washington Wildflowers* (repr. Seattle: Seattle Audubon Society, 1974).

p. 171 *Leslie L. Haskin describes this phenomenon* Haskin, *Wild Flowers of the Pacific Coast,* vii.

22 THE PRESERVATION OF SMALL THINGS

Epigraph: Aldo Leopold, "The Farmer as a Conservationist," 1939; reprinted in Aldo Leopold, *For the Health of the Land: Previously Unpublished Essays and Other Writings,* ed. J. Baird Callicott and Eric T. Freyfogle (Washington, DC: Island Press/Shearwater Books, 1999).

p. 174 *The more threatened the animal pollinator* Heinz Center, "The State of the Nation's Ecosystems," at www.heinzctr.org/ecosystems.

p. 175 *One example is peppermint candy flower* Janice J. Schofield, *Discovering Wild Plants* (Portland: Alaska Northwest Books, 1989), 206.

p. 175 *At the same time, with all its legendary beauty* Conservation International, www.conservation.org.

p. 177 *Jeffrey Caldwell, landscaper-ecologist* Jeffrey Caldwell, personal communication, 2004.

p. 180 *The appellation "DFCLT"* *The Jepson Manual: Higher Plants of California,* ed. James C. Hickman (Berkeley: University of California Press, 1993).

p. 180 *"Here on this hot slope"* Lester Rowntree, "On a Trail of Beauty," in *The Sierra Nevada, the Range of Light,* ed. Roderick Peattie (New York: Vanguard Press, 1947), 98.

p. 181 *"just as the greens are quieting"* Lester Rowntree, *Hardy Californians* (1936; new, expanded ed., Berkeley: University of California Press, 2006), 232.

p. 182 *Umchachich-da* M. Kat Anderson, *Tending the Wild: Native American Knowledge and the Management of California's Natural Resources* (Berkeley: University of California Press, 2005), 60.

23 THE POLLINATORS OF SMALL THINGS

Epigraphs: Jo Brewer, *Wings in the Meadow* (Boston: Houghton Mifflin, 1967); Mike Koslosky, "What's the Buzz? Native Bees," *BayNature Supplement: Gardening for Wildlife with Native Plants,* January–March 2003, 28; Joanne Lauck, *The Voice of the Infinite in the Small: Revisioning the Insect-Human Connection* (Mill Valley, CA: Swan-Raven, 1998), 153.

p. 184 *To foster bees* Brian L. Griffin, *The Orchard Mason Bee: The Life History, Biology, Propagation, and Use of a North American Native Bee* (Bellingham, WA: Knox Cellars, 1999), 26.

p. 184 *Patches of a single species* Koslosky, "What's the Buzz?"

p. 185 *"who made clay nests"* J. Henri Fabre, *The Mason-Bees* (New York: Dodd, Mead, 1914), 8.

p. 186 *Though the native alkali bee* Stephen L. Buchmann and Gary Nabhan, *The Forgotten Pollinators* (Covelo, CA: Island Press, 1996), 186–87.

p. 186 *and orchardists are familiar* Griffin, *Orchard Mason Bee.*

p. 186 *Providing native bees* Deborah K. Rich, "Abuzz About Bees" and "Time for a New Approach to Crop Pollination," *San Francisco Chronicle,* May 21, 2005, F1, F6.

p. 186 *"Throughout fall, winter, and early spring"* Edward S. Ross, "Bee Bank: The Singular Life of a Solitary Bee," *California Wild* 55, no. 1 (Winter 2002): 27.

p. 186 *They learn that in California* Koslosky, "What's the Buzz?" 28.

p. 187 *Vernal pool wildflowers* Robbin W. Thorp and Joan M. Leong, "Specialist Bee Pollinators of Showy Vernal Pool Flowers," in *Ecology, Conservation, and Management of Vernal Pool Ecosystems—Proceedings from a 1996 Conference* (Sacramento, CA: California Native Plant Society, 1998), 169. Other showy wildflowers associated with vernal ponds (and with specialist bee pollinators), that could be good garden subjects include yellow carpet *(Blennosperma),* goldfields *(Lasthenia),* and skyblue *(Downingia).*

p. 189 *"Elderberry gives us the clapper sticks"* Ben Cunningham-Summerfield, "Traditional Uses and Tending of Elderberry: Flute Making and Playing, Foods, Gathering Ethics and Management Practices" (talk presented to the annual meeting of the Society for Ethnobiology, Davis, CA, spring 2004).

24 ANIMAL ASSISTANTS

p. 190 *In his book* The California Quail A. Starker Leopold, *The California Quail* (Berkeley: University of California Press, 1977), 53.

p. 193 *the Choynumne Indians supplied Jeff Mayfield's family* Thomas Jefferson Mayfield, *Tailholt Tales,* by Frank F. Latta (Santa Cruz, CA: Bear State Books, 1976), 27.

p. 195 *One possible control for Argentine ants* Interest Alert, "California Scientists Find Natural Way to Control Spread of Destructive Argentine Ants," www.interest alert.com.

p. 195 *slow down their spread* Kathleen G. Human and Deborah M. Gordon, "Exploitation and Interference Competition between the Invasive Argentine Ant, *Linepithema humile,* and Native Ant Species," *Oecologia* (Heidelberg: Springer Berlin, 2004), 1.

p. 195 *California horned lizards* Insecta Inspecta World, Web site at www .insecta-inspecta.com/ants/argentine/index.html.

p. 195 *Argentine ants like disturbed soil* David A. Holway, "Factors controlling rate of invasion: a natural experiment using Argentine ants," *Oecologia* 115 (1998): 206–12.

26 FOG FLOWER

p. 202 *One writer referred to tarweed* Leslie L. Haskin, *Wild Flowers of the Pacific Coast* (Portland: Binfords and Mort, 1934), 375. "The seeds of this species are rich and oily, and were gathered by the Indians for food. Mrs. Hargreaves writes of the Indian's autumn."

27 SCALE IS ALL

p. 204 *"the bottom line is"* Cass Turnbull, founder of Seattle-based PlantAmnesty, at www.plantamnesty.org, and author of *Cass Turnbull's Guide to Pruning: What, When, Where and How to Prune for a More Beautiful Garden* (Seattle: Sasquatch Books, 2004). See also Valerie Easton, "Tree Torture; When You Lop the Top, You Invite Trouble," *Seattle Times,* December 21, 2003.

p. 204 *Place-names honoring this species* Erwin G. Gudde, *California Place Names,* 4th ed. (Berkeley: University of California Press, 1998), 180.

28 THE QUIET TIME

p. 209 *"the fifth season"* David Rains Wallace, *The Untamed Garden and Other Personal Essays* (Columbus: Ohio State University Press, 1986; repr. New York: Collier Books, 1988), 27–31.

29 THE BERRY HARVEST

p. 211 *"Traditionally, berries were"* Trevor C. Lantz and Nancy J. Turner, "Traditional Phenological Knowledge of Aboriginal Peoples in British Columbia," *Journal of Ethnobiology* 23, no. 2 (2003): 263–86.

p. 211 *These dark purple, red, and deep blue fruits* Julie Beattie, Alan Crozier, and Garry G. Duthie, "Potential Health Benefits of Berries," *Current Nutrition and Food Science* 1 (2005): 71–86.

p. 213 *"so patterned to succor"* Quoted in Skee Hamann's biography, "Edith Van Allen Murphey," 76 (written between 1980–83, it is archived at the Held-Poage Memorial Research Library, Ukiah, CA).

30 PATHS AND TRAILS

Epigraph: Pliny Earle Goddard, *Life and Culture of the Hupa,* University of California Publications in American Archaeology and Ethnology (Berkeley: University of California, 1903), 88.

p. 217 *"[You] lie here still"* Edward Sapir quoted in Richard Keeling, *Cry for Luck: Sacred Song and Speech among the Yurok, Hupa, and Karok Indians of Northwestern California* (Berkeley: University of California Press, 1992), 41.

p. 218 *"[She] wondered what their life had been"* Effie McAbee Hulbert, *Indian Summer,* Anderson Valley Historical Society (Cloverdale, CA: Laurelwood Publishing, 1988), 98.

31 EVERYTHING'S HERE

Epigraphs: Geoffrey Geupel, personal communication, 2004; Stephen W. Edwards, "Learning How to Have a Pond," *Manzanita* 6, no. 1 (Spring 2004).

p. 220 *The California red-legged frog* Information on the red-legged frog is from the U.S. Fish and Wildlife Service, Sacramento, CA.

p. 221 *"To the unmagnified eye"* Eugenia McNaughton (environmental scientist, Environmental Protection Agency), interview, March 2006.

p. 221 *"a writer most fails"* William Least Heat-Moon, *PrairyErth: A Deep Map* (Boston: Houghton Mifflin, 1991), 238.

p. 222 *The leaves, seeds, and twigs* Beatrice M. Beck and Sandra S. Strike, *Ethnobotany of the California Indians,* vol. 2, *Aboriginal Uses of California's Indigenous Plants* (Champaign, IL: Koeltz Scientific Books, 1994), 35.

32 LIFE IS MAINTENANCE

Epigraphs: Henry David Thoreau, quoted in the introduction to *Wild Fruits: Thoreau's Last Discovered Manuscript,* edited and introduced by Bradley P. Dean (New York: W. W. Norton, 2000), xv; Robert Bly, foreword to *Secrets of the Talking Jaguar,* by Martin Prechtel (New York: Jeremy P. Tarcher, 1998), xi.

p. 224 *"The human mind"* Paul Shepard, *The Others: How Animals Made Us Human* (Covelo, CA: Island Press, 1996), 25.

p. 224 *As is well documented by M. Kat Anderson* M. Kat Anderson and Thomas C. Blackburn, eds., *Before the Wilderness: Environmental Management by Native Californians* (Menlo Park, CA: Ballena Press, 1993); and M. Kat Anderson, *Tending the Wild: Native American Knowledge and the Management of California's Natural Resources* (Berkeley: University of California Press, 2005).

CODA WE ARE NOT THE FIRST

Epigraph: Wendell Berry, quoted in Mark Engler, "Go Ahead Mr. Wendell," *Grist Magazine,* August 5, 2004, found on www.grist.org/news/maindish/2004/08/05/engler-berry.

p. 230 *A Yurok story of long ago* A. L. Kroeber, *Yurok Myths* (Berkeley: University of California Press, 1976), 199.

p. 230 *"When he arrived, he saw a bush"* Deerbrush, or buckbrush, probably refers to *Ceanothus integerrimus.*

p. 231 *"In a surprising number"* Aldo Leopold, "The Farmer as a Conservationist," 1939; reprinted in Aldo Leopold, *For the Health of the Land: Previously Unpublished Essays and Other Writings,* ed. J. Baird Callicott and Eric T. Freyfogle (Covelo, CA: Island Press/Shearwater Books, 1999).

p. 232 *"Beauty often comes to us"* Elaine Scarry, "Beauty of the Scholar's Duty to Justice," *Profession,* 2000.

RECOMMENDED READING

Anderson, M. Kat. *Tending the Wild: Native American Knowledge and the Management of California's Natural Resources.* Berkeley: University of California Press, 2005.

Anderson, M. Kat, and Thomas C. Blackburn, eds. *Before the Wilderness: Environmental Management by Native Californians.* Menlo Park, CA: Ballena Press, 1993.

Barrows, David Prescott. *Ethno-Botany of the Coahuilla Indians of Southern California.* 1900. Reprint. Banning, CA: Malki Museum Press, 1977.

Beck, Beatrice M., and Sandra S. Strike. *Ethnobotany of the California Indians.* Vol. 2, *Aboriginal Uses of California's Indigenous Plants.* Champaign, IL: Koeltz Scientific Books, 1994.

Bibby, Brian. *Deeper than Gold: Indian Life in the Sierra Foothills.* Berkeley: Heyday Books, 2005.

The Birds of North America: Life Histories for the 21st Century. Cornell Lab of Ornithology Web site, http://bna.birds.cornell.eduBNA.

Bornstein, Carol, David Fross, and Bart O'Brien. *California Native Plants for the Garden.* Los Olivos, CA: Cachuma Press, 2005.

Bossard, Carla C., John M. Randall, and Marc C. Hoshovsky, eds. *Invasive Plants of California's Wildlands.* Berkeley: University of California Press, 2000.

Brown, Tom. *Tom Brown's Guide to Wild Edible and Medicinal Plants.* New York: Penguin, 1986.

Bruchac, Joseph. *Roots of Survival: Native American Storytelling and the Sacred.* Golden, CO: Fulcrum Publishing, 1996.

Buchmann, Stephen L., and Gary Nabhan. *The Forgotten Pollinators.* Covelo, CA: Island Press, 1996.

Buhner, Stephen. *The Lost Language of Plants: The Ecological Importance of Plant Medicines to Life on Earth.* White River Junction, VT: Chelsea Green Publishing, 2002.

Campbell, Paul D. *Survival Skills of Native California.* Salt Lake City: Gibbs-Smith, 1999.

Chase, J. Smeaton. *California Coast Trails: A Horseback Adventure from Mexico to Oregon.* 1913. Reprint. Palo Alto, CA: Tioga Publishing, 1987.

Dawson, William Leon. *The Birds of California.* San Diego: South Moulton, 1923.

Eisner, Thomas. *For Love of Insects.* Cambridge, MA: Harvard University Press, Belknap Press.

Fabre, J. Henri. *The Mason-Bees.* New York: Dodd, Mead, 1914.
Gayton, Anna H. *Yokuts and Western Mono Ethnography.* Anthropological Records, vol. 10, no. 1. Berkeley: University of California, 1948.
Gerda Isenberg: California Native Plants Nurserywoman, Civil Rights Activist, and Humanitarian. Interviews conducted by Suzanne B. Reiss, with an introduction by Bart O'Brien. Regional Oral History Office, Bancroft Library, University of California, Berkeley, 1991.
Goddard, Pliny Earle. *Life and Culture of the Hupa.* University of California Publications in American Archaeology and Ethnology. Berkeley: University of California Press, 1903.
Gudde, Erwin G. *California Place Names.* 4th ed. Berkeley: University of California Press, 1998.
Haskin, Leslie L. *Wild Flowers of the Pacific Coast.* Portland: Binfords and Mort, 1934.
Hickman, James C., ed. *The Jepson Manual: Higher Plants of California.* Berkeley: University of California Press, 1993.
Hulbert, Effie McAbee. *Indian Summer.* Anderson Valley Historical Society. Cloverdale, CA: Laurelwood Publishing, 1988.
Jordan, William R. III. *The Sunflower Forest.* Berkeley: University of California Press, 2003.
Keeling, Richard. *Cry for Luck: Sacred Song and Speech among the Yurok, Hupa, and Karok Indians of Northwestern California.* Berkeley: University of California Press, 1992.
Kelly, Isabel. *Interviews with Tom Smith and Maria Copa: Isabel Kelly's Ethnographic Notes on the Coast Miwok Indians of Marin and Southern Sonoma Counties.* Edited by Mary E. T. Collier and Sylvia Barker Thalman. San Rafael, CA: Miwok Archaeological Preserve of Marin, 1991.
Kozloff, Eugene N., and Linda H. Beidleman. *Plants of the San Francisco Bay Region: Mendocino to Monterey.* Pacific Grove, CA: Sagen Press, 1994.
Kreitlinger, Kim, and Tom Gardali. *Bringing the Birds Back: A Guide to Habitat Enhancement in Riparian and Oak Woodlands for the North Bay Region.* California Partners in Flight Regional Bird Conservation Plan no. 1, www.prbo.org/calpif. 2006.
Kroeber, A. L. *Yurok Myths.* Berkeley: University of California Press, 1976.
Laird, Carobeth. *Mirror and Pattern.* Banning, CA: Malki Museum Press, 1984.
———. *The Chemehuevis.* Banning, CA: Malki Museum Press, 1976.
Lanner, Ronald. *Made for Each Other: A Symbiosis of Birds and Pines.* New York: Oxford University Press, 1996.
Larrison, Earl J., Grace W. Patrick, William H. Baker, and James A. Yaich. *Washington Wildflowers.* Reprint. Seattle: Seattle Audubon Society, 1974.

Latta, Frank F. *Handbook of Yokuts Indians.* Santa Cruz: Bear State Books, 1977.

Least Heat-Moon, William. *PrairyErth: A Deep Map.* Boston: Houghton Mifflin, 1991.

Lee, Gaylen D. *Walking Where We Lived: Memoirs of a Mono Indian Family.* Norman: Oklahoma University Press, 1998.

Lee, Milicent. *Indians of the Oaks.* 1937. Reprinted Ramona, CA: Acoma Books, 1978.

Leopold, A. Starker. *The California Quail.* Berkeley: University of California Press, 1977.

Leopold, Aldo. *For the Health of the Land: Previously Unpublished Essays and Other Writings.* Edited by J. Baird Callicott and Eric T. Freyfogle. Washington, DC: Island Press/Shearwater Books, 1999.

Luthin, Herbert. *Surviving through the Days: Translations of Native California Stories and Songs.* Berkeley: University of California Press, 2002.

Maser, Chris, and James Trappe, eds. *The Seen and Unseen World of the Fallen Tree,* General technical report PNW-164. Portland: Pacific Northwest Forest and Range Experiment Station, U.S. Department of Agriculture, Forest Service, 1984.

Maser, Chris, Robert F. Tarrant, James M. Trappe, and Jerry F. Franklin, eds. *From the Forest to the Sea: A Story of Fallen Trees.* General technical report PNW-GTR-299. Portland: Pacific Northwest Forest and Range Experiment Station, U.S. Department of Agriculture, Forest Service, 1988.

Mayfield, Thomas Jefferson. *Tailholt Tales,* by Frank F. Latta. Santa Cruz, CA: Bear State Books, 1976.

McKibbin, Grace. *In My Own Words.* Berkeley: Heyday Books, 1997.

Merriam, C. Hart, ed. *The Dawn of the World: Myths and Tales of the Miwok Indians of California.* 1910. Lincoln: University of Nebraska Press, 1993.

Miller, Joaquin. *Life amongst the Modocs: Unwritten History.* 1873. Reprint. San Jose: Union Press, 1982.

Murphey, Edith Van Allen. *Indian Uses of Native Plants.* Ukiah: Mendocino County Historical Society, 1959.

Ortiz, Bev. *It Will Live Forever: Traditional Yosemite Indian Acorn Preparation.* Berkeley: Heyday Press in association with Rick Heide, 1991.

Pollan, Michael. *The Botany of Desire: A Plant's Eye View of the World.* New York: Random House, 2001.

Powers, Stephen. *Tribes of California.* Introduction and notes, Robert F. Heizer. Berkeley: University of California Press, 1976.

Prechtel, Martin. *Secrets of the Talking Jaguar.* Foreword by Robert Bly. New York: Jeremy P. Tarcher, 1998.

Raphael, Ray. *Tree Talk: The People and Politics of Timber.* Washington, DC: Island Press, 1981.

———. *More Tree Talk: The People, Politics, and Economics of Timber.* Washington, DC: Island Press, 1994.

Rowntree, Lester. *Hardy Californians.* 1936. New, expanded edition. Berkeley: University of California Press, 2006.

Sawyer, John O., and Todd Keeler-Wolf, *A Manual of California Vegetation.* Sacramento: Native Plant Society, 1997.

Schofield, Janice J. *Discovering Wild Plants.* Portland: Alaska Northwest Books, 1989.

Shepard, Paul. *The Others: How Animals Made Us Human.* Covelo, CA: Island Press, 1996.

Simon, Chris. *Down an Old Road: The Poetic Life of Wilma Elizabeth McDaniel.* Videocasette. 2001 (available from Chris Simon, csimon@lasal.net).

Stanger, Margaret. *That Quail, Robert.* Philadelphia: Lippincott, 1966.

Stokes, Donald. *A Guide to Observing Insect Lives.* Stokes Nature Guides. Boston: Little, Brown, 1983.

Underwood, Caroline. *The Salmon Forest.* CBC, *The Nature of Things.* CBC Enterprises, 2001.

Wallace, David Rains. *The Untamed Garden and Other Personal Essays.* Columbus: Ohio State University Press, 1986; reprinted New York: Collier Books, 1988.

Welmas, Carolina. *Muluwetam.* Banning, CA: Malki Museum Press, 1973.

INDEX

Designer:	Sandy Drooker
Text:	10.5/14 Adobe Garamond
Display:	Helvetica
Compositor:	BookMatters, Berkeley
Indexer:	Patricia Deminna
Illustrator:	Kathleen O'Neill
Printer and binder:	Sheridan Books, Inc.